Recommender Systems
and the Social Web

Fatih Gedikli

Recommender Systems and the Social Web

Leveraging Tagging Data for Recommender Systems

Springer Vieweg

Dr. Fatih Gedikli
Dortmund, Germany

Dissertation Technische Universität Dortmund, 2012

ISBN 978-3-658-01947-1   ISBN 978-3-658-01948-8 (eBook)
DOI 10.1007/978-3-658-01948-8

The Deutsche Nationalbibliothek lists this publication in the Deutsche Nationalbibliografie; detailed bibliographic data are available in the Internet at http://dnb.d-nb.de.

Library of Congress Control Number: 2013935227

Springer Vieweg
© Springer Fachmedien Wiesbaden 2013

Printed on acid-free paper

Springer Vieweg is a brand of Springer DE.
Springer DE is part of Springer Science+Business Media.
www.springer-vieweg.de

Dedicated To My Family.

Contents

List of Figures

List of Tables

Chapter 1

Introduction

1.1 Motivation

Due to the considerable growth of information available online, it has become a constant challenge to help Internet users to deal with the corresponding information overload. Over the last decade, various techniques in the areas of information retrieval and filtering have been developed to help users find items that match their information needs and filter out unrelated information items [Hanani et al., 2001].

In contrast to information retrieval and filtering techniques implemented in search engines, whose aim is to retrieve the desired information from a large amount of information based on a user query, recommender systems are today commonly in use on e-commerce platforms. They help online visitors find relevant information or items to purchase in a personalized way [Jannach et al., 2010; Ricci et al., 2011a]. In the age of information overload recommender system technologies are of high importance for the success of large-scale e-commerce sites. Business Insider[1], for example, names several recommender system technologies, such as Amazon.com's recommendation engines or Google's news algorithm, among the 11 most essential algorithms that "make the Internet work".

When applied in the context of e-commerce, the aim of recommender systems is to provide personalized recommendations that best suit a customer's taste, preferences, and individual needs [Resnick and Varian, 1997; Adomavicius and Tuzhilin, 2005]. Besides this, a recommender system is supposed to explain the underlying reasons for its proposals to the user. An example for an explanation would be Amazon.com's *"Customers who bought this item also bought..."* label for a recommendation list, which also carries explanatory information. Explanations for recommendations have increasingly gained in importance over the last years both in academia and industry because they can significantly influence the user-perceived quality of such a system [Tintarev and Masthoff, 2007a].

The advantages of using recommender systems are manifold. For example, they can help to build better relationships with customers, increase the value of e-business, or broaden sales diversity [Nikolaeva and Sriram, 2006; Dias et al., 2008; Fleder and Hosanagar, 2009]. In practice, recommender systems have been implemented in different commercial domains such as travel and tourism, entertainment, or book sales, as mentioned in [Ricci and Nguyen, 2007], [Jannach and Hegelich, 2009], or [Linden et al., 2003].

With the advent of the Social Web, user generated content has enriched the social dimension of the Web [Kim et al., 2010b]. New types of Web applications such as Delicious and Flickr[2] have emerged which emphasize content sharing and collaboration. These so-called Social Web platforms turned users from passive recipients of information into active and engaged contributors. As a result, the amount of user contributed information provided by the Social Web poses both new possibilities and challenges for recommender system research [Freyne et al., 2011].

This thesis focuses on the challenging topic of leveraging these new sources of knowledge to enhance existing recommender system techniques and introduce new recommendation approaches based on Social Web data. In particular, we focus on tagging data and propose new ways to leverage user-contributed tags in recommender systems.

---

[1]http://www.businessinsider.com/internet-algorithms-2011-8
[2]http://www.delicious.com, http://www.flickr.com

1.2   Concept of rating items by rating tags

User-contributed tags are today a popular means for users to organize and retrieve items of interest in the Social Web [Vig et al., 2009]. A tag is simply a user-contributed keyword or phrase which can be assigned to resources[3] such as books, CDs, movies, and news. They are used to convey meta-information about the resource they are assigned to.

Social Tagging plays an increasingly important role both on Social Web platforms such as last.fm and YouTube[4] as well as on large-scale e-commerce sites such as Amazon.com. Social Web applications encourage users to share and collaboratively classify content using tags.

This thesis deals with questions of how tagging data can be exploited by recommender systems in the best possible way to improve both the quality of recommendations as well as the quality of the corresponding explanations. Recommendation quality is usually measured by accuracy metrics such as the mean absolute error (MAE) and the root mean squared error (RMSE), whereas different evaluation factors exist for explanation quality such as efficiency and effectiveness. We will introduce the different evaluation metrics and factors later on in this work. We will also show ways to improve the limited quality of user-contributed tags with the help of recommender systems. Throughout this thesis, we will look at the whole recommender ecosystem, that is, leveraging tagging data for recommender systems and vice versa. This ecosystem is visualized in Figure 1.1 for a movie recommendation scenario.

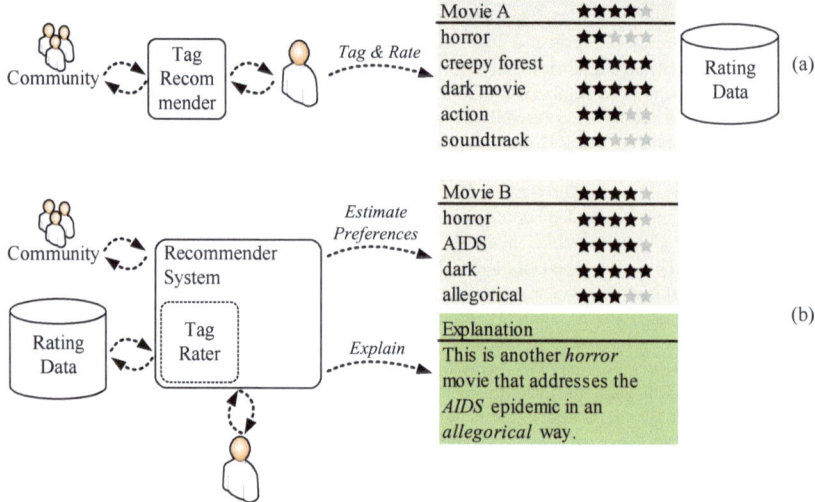

Figure 1.1: (a) Tagging and rating items. (b) Recommending items and explaining recommendations by exploiting tag preference data.

Figure 1.1 (a) shows the process of preference acquisition. Recommender systems usually base their recommendations on an implicitly or explicitly acquired user profile which is the system's representation of the user's interests and characteristics. In this work, we propose to extend the usual user-item rating matrix not only by a set of user-provided tags for the items, but also by *tag preferences* describing the user's opinion about the item features represented by these tags. Thus, we are able to build a more detailed profile about the user and his or her interests. In Figure 1.1 (a) the user assigns or chooses one or more tags for the item to be rated. The user can either create new tags or select existing quality tags in the sense of [Sen et al., 2007] from a list of recommended tags, which can be provided by a tag recommender system such as the ones proposed in [Jäschke et al., 2007] or [Zhang et al., 2009a]. After

---

[3]According to the W3C, a resource can be any entity identified by a uniform resource identifier.
[4]http://www.last.fm, http://www.youtube.com

assigning a tag, each individual tag can be given a rating, that is, the user rates selected tags representing the item features and assigns an overall rating to the item.

Figure 1.1 (b), on the other hand, shows scenarios where the augmented rating matrix is exploited by a recommender system. The recommender uses, amongst others, tag preference data to estimate a user's preference value for a given item. Tag preferences can either be acquired explicitly or derived automatically using the "Tag Rater" component in Figure 1.1 (b) in case no such explicit information is available. Detailed tag preferences can, for example, be estimated from the items' overall ratings. Figure 1.1 (b) also provides another compelling application scenario in which tag preferences are used for explaining recommendations. Next, we will briefly discuss the main contributions of the thesis.

1.3 Contributions

Tag-based recommendations and explanations are the two main areas of contribution in this work. The area of tag-based recommendations deals mainly with the topic of recommending items by exploiting tagging data. It covers the process depicted in Figure 1.1 (a) and partly the process depicted in Figure 1.1 (b) (the "Estimate Preferences" link):

- **Tag recommendation algorithm.** Tag recommenders are designed to help the online user in the tagging process and suggest appropriate tags for resources with the purpose to increase the tagging quality. In this thesis we propose a tag recommender algorithm called *LocalRank* which can generate highly-accurate tag recommendations in real-time.

- **Concept of user- and item-specific tag preferences.** We propose the concept of user- and item-specific tag preferences in this work. By "attaching feelings to tags" users are provided a powerful means to express in detail which features of an item they particularly like or dislike. When following such an approach, users would therefore not only add tags to an item as in usual Web 2.0 applications, but also attach a preference to the tag itself, expressing, for example, whether or not they liked a certain actor in a given movie.

- **Recommendation schemes that take tag preferences into account.** The introduction of user- and item-specific tag preferences lead to the development of new recommendation schemes that can exploit tag preference data. In this work, we present first algorithms that consider tag preferences to generate more accurate predictions. Note that often explicit tag preference data is not available or the data is very sparse. Therefore, we additionally propose a metric to infer context-specific tag preferences automatically (see the "Tag Rater" component in Figure 1.1 (b)). The evaluation on two different data sets reveals that our recommendation scheme is capable of providing more accurate recommendations than previous tag-based recommender algorithms and recent tag-agnostic matrix factorization techniques.

The area of tag-based explanations deals with questions of how explanations for recommendations should be communicated to the user in the best possible way. It covers the "Explain" link presented in Figure 1.1 (b):

- **Explanations based on tag clouds using tag preferences.** We present the results of a user study in which three explanation approaches are evaluated with respect to the desired effects and quality dimensions *efficiency, effectiveness & persuasiveness, satisfaction,* and *trust*. We compare keyword-style explanations, which performed best according to effectiveness in previous work, with two new explanation methods based on personalized and non-personalized tag clouds. The personalized tag cloud interface makes use of the idea of user- and item-specific tag preferences described above. The results show that users can make better decisions faster when using the tag cloud interfaces rather than the keyword-style explanations. In addition, users generally favored the tag cloud interfaces over keyword-style explanations.

- **Analysis of the effects of using different explanation styles.** We present the results of another user study in which ten explanation approaches are evaluated. We compare the tag cloud explanations from our first study with other explanation interfaces from previous work. In this study, we additionally analyze their effects on the quality dimension *transparency*. The results show that, in particular, the newly proposed tag cloud interfaces represent examples for explanation interfaces which can help increase the user's trust in a system. They are effective, transparent, and improve user satisfaction. Besides an analysis of different effects, we provide a first analysis of the interdependencies between the different evaluation factors of explanations.

- **Design guidelines.** Based on the insights and other observations of both user studies, we derive a first set of possible guidelines for designing or choosing an explanation interface for a recommender system. We see our set of design guidelines as a helpful tool for researchers as well as practitioners.

1.4   Overview of the dissertation

The thesis is organized as follows:

- **Chapter 2** presents the basic concepts of recommender systems. Recommendation techniques where the task is to predict user preferences for items the user has not seen before are introduced. These prediction values can be used to compute "top-n" recommendation lists. The chapter also provides an overview of different types of evaluation procedures commonly found in the recommender system literature. Finally, we present current recommendation approaches based on Social Web data. In particular, a selection of recent developments in the field of tag-based recommender systems is presented.

- **Chapter 3** describes a tag recommending algorithm called LocalRank. The algorithm is inspired by the state-of-the-art algorithm FolkRank. In contrast to FolkRank the rank weights in LocalRank are based only on the local "neighborhood" of a given user and resource. Therefore, LocalRank only considers a small part of the user-resource-tag graph and can generate highly-accurate tag recommendations in real-time. The algorithms are evaluated on a popular social bookmarking data set.

- **Chapter 4** introduces the concept of user- and item-specific tag preferences and explores algorithms that can exploit tag preference data to improve the predictive accuracy of recommender systems. New schemes to infer and exploit context-specific tag preferences in the recommendation process are presented. The approaches are evaluated on two different data sets from the movie domain.

- **Chapter 5** describes the results of our first user study, in which we analyze users' reactions to three different explanation interfaces for collaborative filtering recommender systems. We compare keyword-style explanations with two new explanation methods based on personalized and non-personalized tag clouds. The personalized tag cloud interface makes use of the idea of tag preferences described in Chapter 4.

- **Chapter 6** discusses the results of our second user study, in which we analyze users' reactions to ten different explanation styles. Besides an analysis of different effects, a first analysis of the interdependencies between the different aims of explanations is provided. Finally, a set of possible guidelines for designing or choosing explanations is introduced.

- Lastly, in **Chapter 7** the results and the key contributions of the thesis as well as the limitations are summarized. The thesis concludes with an outlook on future work.

1.5   Publications

Almost all of the results presented in this thesis have already been published in reputable peer-reviewed international journals and conferences or are at the stage of being reviewed. Details about the individual contributions can be found in Appendix A.

- Marius Kubatz, Fatih Gedikli, Dietmar Jannach: LocalRank - Neighborhood-based, fast computation of tag recommendations, *Proceedings of the 12th International Conference on Electronic*

Commerce and Web Technologies (EC-Web'11), Toulouse, France, 2011, pp. 258-269.
The content of Chapter 3 is based on this publication in the proceedings of a peer-reviewed international conference.

- Fatih Gedikli, Dietmar Jannach: Improving recommendation accuracy based on item-specific tag preferences, *ACM Transactions on Intelligent Systems and Technology (ACM TIST)*, Volume 4, Number 1, 2013.
  Chapter 4 is build on this paper from a peer-reviewed international journal.

- Fatih Gedikli, Mouzhi Ge, Dietmar Jannach: Understanding recommendations by reading the clouds, *Proceedings of the 12th International Conference on Electronic Commerce and Web Technologies (EC-Web'11)*, Toulouse, France, 2011, pp. 196-208.
  This peer-reviewed international conference paper builds the basis for Chapter 5.

- Fatih Gedikli, Dietmar Jannach, Mouzhi Ge: Effects of recommender system explanations on users - An analysis.
  The results of this paper, which is currently being reviewed, are depicted in Chapter 6.

Note that several other papers which build the basis for the main papers listed above or are related to recommender system research but not covered in this thesis were published by the author of this work:

- Mouzhi Ge, Dietmar Jannach, Fatih Gedikli, Martin Hepp: *Effects of the placement of diverse items in recommendation lists, Proceedings of the 14th International Conference on Enterprise Information Systems (ICEIS'12)*, Wroclaw, Poland, 2012, pp. 201-208.

- Dietmar Jannach, Zeynep Karakaya, Fatih Gedikli: Accuracy improvements for multi-criteria recommender systems, *Proceedings of the ACM Conference on Electronic Commerce (EC'12)*, Valencia, Spain, 2012, pp. 674-689.

- Dietmar Jannach, Fatih Gedikli, Zeynep Karakaya, Oliver Juwig: Recommending hotels based on multi-dimensional customer ratings, *Proceedings of the 19th eTourism Community Conference on eTourism Present and Future Services and Applications (ENTER'12)*, Helsingborg, Sweden, 2012, pp. 320-331.

- Fatih Gedikli, Faruk Bagdat, Mouzhi Ge, Dietmar Jannach: RF-Rec: Fast and accurate computation of recommendations based on rating frequencies, *Proceedings of the 13th IEEE Conference on Commerce and Enterprise Computing (CEC'11)*, Luxembourg City, Luxembourg, 2011, pp. 50-57.

- Fatih Gedikli, Mouzhi Ge, Dietmar Jannach: Explaining online recommendations using personalized tag clouds, *i-com Journal* (http://i-com-media.de), 2011, pp. 3-10.

- Mouzhi Ge, Fatih Gedikli, Dietmar Jannach: Placing high-diversity items in top-n recommendation lists, *Proceedings of the 9th Workshop on Intelligent Techniques for Web Personalization & Recommender Systems at IJCAI'11*, Barcelona, Spain, 2011, pp. 65-68.

- Fatih Gedikli, Dietmar Jannach: Neighborhood-restricted mining and weighted application of association rules for recommenders, *Proceedings of the 11th International Conference on Web Information System Engineering (WISE'10)*, Hong Kong, China, 2010, pp. 157-165.

- Fatih Gedikli, Dietmar Jannach: Recommending based on rating frequencies, *Proceedings of the 4th ACM Conference on Recommender Systems (RecSys'10)*, Barcelona, Spain, 2010, pp. 233-236.

- Fatih Gedikli, Dietmar Jannach: Rating items by rating tags, *Proceedings of the 2nd Workshop on Recommender Systems and the Social Web at ACM RecSys'10*, Barcelona, Spain, 2010, pp. 25-32.

- Fatih Gedikli, Dietmar Jannach: Neighborhood-restricted mining and weighted application of association rules for recommenders, *Proceedings of the 8th Workshop on Intelligent Techniques for Web Personalization & Recommender Systems at UMAP'10*, Big Island, Hawaii, 2010, pp. 8-19.

- Fatih Gedikli, Dietmar Jannach: Recommending based on rating frequencies: Accurate enough?, *Proceedings of the 8th Workshop on Intelligent Techniques for Web Personalization & Recommender Systems at UMAP'10*, Big Island, Hawaii, 2010, pp. 65-70.

Lastly, two other papers were co-authored by the author of this thesis before the time of the doctoral studies but are not related to recommender system research at all:

- Paul Lokuciejewski, Fatih Gedikli, Peter Marwedel: Accelerating WCET-driven optimizations by the invariant path - A case study of loop unswitching, *Proceedings of the 12th International Workshop on Software and Compilers for Embedded Systems at DATE'09*, Nice, France, 2009, pp. 11-20.

- Paul Lokuciejewski, Fatih Gedikli, Peter Marwedel, Katharina Morik: Automatic WCET reduction by machine learning based heuristics for function inlining, *Proceedings of the 3rd Workshop on Statistical and Machine Learning Approaches to Architecture and Compilation at HiPEAC'09*, Paphos, Cyprus, 2009, pp. 1-15.

Chapter 2

Preliminaries

This chapter is organized as follows: First we present the two basic filtering techniques for recommender systems: collaborative and content-based filtering. Afterwards, in Section 2.2, we discuss how recommender systems can be compared according to different quality aspects that are relevant for the application. We describe the different types of evaluation procedures commonly found in the recommender system literature and their respective application domains. Finally, in Section 2.3, we present current recommendation approaches based on Social Web data. We focus on tagging data and identify the different approaches to leverage tagging data in recommender systems.

2.1 Basic recommendation techniques

Although recommender systems have their roots in information retrieval [Hanani et al., 2001], from the mid-1990s recommender systems have become an independent research area of their own, see [Adomavicius and Tuzhilin, 2005; Jannach et al., 2010] or [Ricci et al., 2011a] for recent overviews. Commonly, recommender systems are classified into the categories collaborative filtering, content-based filtering, and hybrid approaches [Melville and Sindhwani, 2010].

Collaborative filtering approaches exploit the wisdom of the crowd and recommend items based on the similarity of tastes or preferences of a larger user community. *Content-based* approaches, on the other hand, recommend items by analyzing their features to identify those items that are similar to the ones that the user preferred in the past. Besides to that, *knowledge-based* recommender systems exist in the literature (see, for example, [Felfernig and Burke, 2008]), which rely on explicit user requirements and some form of means-ends knowledge to match the user's needs with item characteristics, but are not covered in this thesis. In order to benefit from the advantages of the different main approaches, *hybrid* recommender systems try to combine different algorithms and exploit information from various knowledge sources. Studies have shown that for example hybrids which combine content-based and collaborative filtering can lead to more accurate predictions than pure collaborative filtering or content-based recommenders [Balabanovic and Shoham, 1997b; Melville et al., 2002].

2.1.1 Notation and symbols

In the following, we will introduce a notation and symbols for defining the recommendation problem more precisely. Our subsequent definition of the recommendation problem is based on the definition of [Adomavicius and Tuzhilin, 2005].

Let $U = \{u_1, ..., u_n\}$ be the set of users and let $I = \{i_1, ..., i_m\}$ be the set of items that can be recommended to the users. Furthermore we assume that $\hat{r} : U \times I \to S$ is a utility function which measures the usefulness $\hat{r}_{u,i}$ of item i to user u and returns a ranking in a totally ordered set S consisting of real numbers or nonnegative integers. Note that in the recommender system literature the utility of an item is usually represented by a *rating* value which stands for the degree a particular user likes a given item. Then, according to [Adomavicius and Tuzhilin, 2005], the recommendation problem consists of selecting for each user $u \in U$ a not-yet-rated item $i'_u \in I$ that maximizes the utility function $\hat{r}$:

$$\forall u \in U, \quad i'_u = \arg\max_{i \in I} \hat{r}_{u,i} \tag{2.1}$$

Since the utility function is used to *predict* a user's interest in a particular item, we will regard $\hat{r}$ as the prediction function in the following. Therefore, $\hat{r}_{u,i}$ stands for the predicted rating value of user u for item i. In this context, we call user u and item i *target user* and *target item* respectively. Together they build the user and item pair for which a rating prediction is made. The target user is sometimes also called the *active user*. Note that the prediction function is usually estimated (learned) in many different ways, e.g., by using methods from machine learning. The basic underlying assumption is that user ratings from the past must also be predictive of the future.

The customer ratings can be organized in a $n \times m$ rating matrix R where $r_{u,i}$ represents the rating value of user u for item i, see Figure 2.1. Note that in practice the rating matrix R is often very sparse because usually users only provide very few ratings. We use $\overline{r_u}$ and $\overline{r_i}$ to denote the average rating value of user u and item i respectively. The ratings from a given user u can be viewed as an incomplete rating vector $\vec{u}$. Analogously, $\vec{i}$ represents the incomplete rating vector which holds the rating values assigned to the item i.

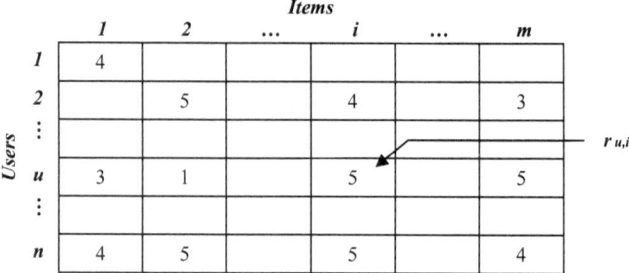

Figure 2.1: The $n \times m$ rating matrix R where $r_{u,i}$ corresponds to the rating value (5) of user u for item i. In this example, the rating values in the user-item rating-matrix range from 1 (strongly dislike) to 5 (strongly like). The main task of a recommender is to predict the missing rating values.

2.1.2 Collaborative filtering

From the mentioned categories above, collaborative filtering is considered to be one of the most successful and promising technologies in practice [Goldberg et al., 1992; Konstan et al., 1997; Sarwar et al., 2000; Adomavicius and Tuzhilin, 2005; Jannach et al., 2010].

We will briefly describe the basic idea of collaborative filtering with the aid of an example. Table 2.2 shows an example user-item rating matrix. For simplicity, we only use a binary "Like/Dislike" rating scale. The task is to predict whether *Alice* will enjoy the movie *Scarface* or not.

	Heat (1995)	Scarface (1983)	Amélie (2001)	Eat Pray Love (2010)
Alice	Dislike	?	Like	Like
User2	Like		Dislike	Dislike
User3	Dislike	Dislike	Like	
User4		Dislike	Like	Like

Figure 2.2: Example user-item rating matrix.

In Table 2.2 we see that the users *User3* and *User4* are "similar" to *Alice* with regard to the provided ratings. Similar users are often referred to as peer users or nearest neighbors in the literature. Since both users provided a *Dislike* statement for the target movie *Scarface*, a collaborative filtering technique will predict that *Alice* will also dislike this movie. Therefore, the movie *Scarface* is not recommended to *Alice*. Thus, collaborative filtering techniques exploit the user preferences provided by the community, that is, the wisdom of the crowd, to make a recommendation.

Memory-based vs. model-based

Collaborative filtering techniques can be further classified as memory-based or model-based [Su and Khoshgoftaar, 2009; Jannach et al., 2010]. In a *memory-based* approach the whole user-item rating matrix is kept in memory and directly used for computing predictions. In Chapter 3 a memory-based implementation is presented for our tag recommender algorithm LocalRank. In a *model-based* approach, on the other hand, the user-item rating matrix is used as input to *learn* a prediction model which is then used at run-time to make recommendations. Such a model can, for example, be the result of a data mining or machine learning algorithm. In Chapter 4, for example, a model-based method using Support Vector Machines (SVM) is introduced. For a comprehensive survey of memory-based and model-based collaborative filtering techniques the reader is referred to [Su and Khoshgoftaar, 2009].

Recommendation scheme

A traditional collaborative filtering algorithm takes a user-item rating matrix as the only input and provides a prediction function that estimates a preference $\hat{r}_{u,i}$ of user u for item i the user has not experienced before. These prediction values can then be used to compute "top-n" recommendation lists.

Usually a user's feeling about an item is encoded as a rating value on a seven-point or ten-point Likert response scale which ranges from "strongly like" to "strongly dislike". Note that user preferences[1], which build the basis for the system's representation of the user's preferences, interests, or characteristics, can be acquired explicitly or implicitly [Miller et al., 2004]. Explicitly acquired preference values represent explicit statements by the users, whereas implicitly acquired preference values are derived automatically, e.g., by monitoring user behavior. Product ratings provided by users on Web sites such as MovieLens[2] or Amazon.com are examples for explicit ratings, whereas product views or the time spent on a Web page are examples for implicit ratings.

Recommender system research in the past has mainly focused on the task of exploiting the user-item rating matrix R to learn a prediction function that accurately estimates a user's real preference values. In Section 2.2.1, we present accuracy metrics that can be used to measure the quality of the prediction function of a recommender.

In collaborative filtering, the prediction value $\hat{r}_{u,i}$ is calculated by exploiting the preferences of a larger user community, that is, the wisdom of the crowd. A basic collaborative filtering recommendation scheme for computing a prediction could work as follows:

1. *Neighborhood formation:* First, the neighbors of the target user u are determined. The neighbors represent a set of like-minded users who share similar tastes with the target user.

2. *Neighborhood selection:* Second, from the set of all neighbors the k nearest neighbors of the target user u are selected. Note that only neighbors are taken into account who provided a rating for the target item i.

3. *Aggregation of ratings:* Lastly, the nearest neighbors' ratings for the target item are aggregated in the final prediction value $\hat{r}_{u,i}$. Aggregation can, for example, be accomplished by building the average rating value over the neighbors' ratings for the target item.

If these steps are repeated for all items the user has not seen before, a top-n recommendation list can be generated by ranking the products according to their estimated rating values. It is important to recall that the basic underlying assumption of this recommendation scheme is that users who shared similar tastes in the past, will share similar tastes in the future. Next, we will discuss these steps in more detail.

---

[1]Throughout this work, we will use the terms "rating" and "preference" interchangeably. The same holds for "item" and "product".
[2]http://www.movielens.org

Neighborhood formation

Collaborative filtering approaches are usually based on neighborhood models due to their intuitiveness and simplicity. A user-based neighborhood model covers the relationships between users (see, for example, [Resnick et al., 1994]), whereas an item-based neighborhood model captures the relationships between items (see, for example, [Sarwar et al., 2001]). In the recommendation scheme described above the preferences from a user-based neighborhood were used to compute the final rating prediction. In order to find similar users or items, a similarity metric has to be defined. Pearson's correlation coefficient is a commonly used measure in recommender systems to compute the similarity between two users a and b [Jannach et al., 2010] and is defined as follows:

$$sim(a,b) = \frac{\sum_{i \in I}(r_{a,i} - \overline{r_a})(r_{b,i} - \overline{r_b})}{\sqrt{\sum_{i \in I}(r_{a,i} - \overline{r_a})^2}\sqrt{\sum_{i \in I}(r_{b,i} - \overline{r_b})^2}} \tag{2.2}$$

Generally speaking, the Pearson correlation coefficient measures the linear dependence between two variables and returns a value between -1 and 1 inclusive. Negative values correspond to a negative correlation (low similarity), while positive values indicate a positive correlation (high similarity). When calculating the correlation between users, the Pearson coefficient also accounts for a user's individual rating behavior by taking the user's average rating into account. Thus, a normalization is achieved before combining the rating vectors of user a and b. Note that the sum in Equation (2.2) only iterates over items $i \in I$ for which both users have provided a rating. However, in practice, extensions to the basic approaches are applied to improve performance because data sets which can often be found in practice are very sparse. *Default voting* describes one such extension which assumes a default rating value, e.g., the user's average rating, for items for which one of the users has not provided an explicit rating [Breese et al., 1998].

Besides the Pearson correlation coefficient, other correlation coefficients such as cosine similarity and adjusted cosine similarity exist in the literature [Salton, 1989; Herlocker et al., 2004; Adomavicius and Tuzhilin, 2005; Jannach et al., 2010]. The pure cosine similarity measure is formally defined as follows:

$$sim(\vec{a},\vec{b}) = cos\angle(\vec{a},\vec{b}) = \frac{\vec{a} \cdot \vec{b}}{|\vec{a}| * |\vec{b}|} \tag{2.3}$$

This measure computes the similarity between two rating vectors $\vec{a}$ and $\vec{b}$ and returns values between 0 (for totally orthogonal vectors) and 1 (for vectors pointing in the same direction), where higher values indicate a strong similarity. A drawback of this similarity measure is that, unlike the Pearson coefficient, it does not take the average rating behavior of the users into account. This motivated the introduction of the adjusted cosine similarity metric:

$$sim(a,b) = \frac{\sum_{u \in U}(r_{u,a} - \overline{r_u})(r_{u,b} - \overline{r_u})}{\sqrt{\sum_{u \in U}(r_{u,a} - \overline{r_u})^2}\sqrt{\sum_{u \in U}(r_{u,b} - \overline{r_u})^2}} \tag{2.4}$$

The adjusted cosine similarity metric factors out the average rating behavior $\overline{r_u}$ of a user u and returns a similarity value between -1 and 1 inclusive, similar to the Pearson coefficient.

The interesting question here is when to use which similarity metric. Recommender system research has analyzed the effects of different similarity metrics on the prediction accuracy of a recommender algorithm. In [Herlocker et al., 1999], Herlocker et al. suggest to use the Pearson coefficient in a user-based recommendation approach. When using an item-based approach, however, the results of the study in [Sarwar et al., 2001] indicate that the adjusted cosine similarity metric outperforms both the basic cosine as well as the Pearson similarity metric regarding the quality dimension accuracy of a recommender algorithm. The effects of different similarity metrics for items were also analyzed in more recent works [Gedikli and Jannach, 2010c; Ekstrand et al., 2011]. These results confirm the finding of [Sarwar et al., 2001]. Therefore, adjusted cosine similarity metric is often used for computing the proximity between items, whereas Pearson coefficient is appropriate for computing the proximity between users.

Neighborhood selection

The neighborhood size k plays a crucial role for the prediction quality of a nearest neighbor collaborative filtering algorithm [Herlocker et al., 1999; Sarwar et al., 2001]. If k is chosen too large, the risk of increased noise in the data will be high because not all of the neighbors are good "predictors". Additionally, the time required for generating predictions increases. On the other hand, if k is chosen too small, the prediction quality may be affected because the prediction is based only on a few neighbors. A detailed discussion of this tradeoff relation can be found in [Herlocker et al., 1999; Anand and Mobasher, 2005; Jannach et al., 2010].

Different techniques exist to select a neighborhood size. For example, a similarity threshold can be provided such that only neighbors with a higher similarity are taken into account. The similarity threshold has to be selected properly. Otherwise, the neighborhood size could be too small or large leading to the problems discussed above. We suggest to use a user-dependent similarity threshold in order to account for users with different neighborhood sizes. Another possibility is to use a fixed neighborhood size k, leading to the question of which value chosen for k is the best. In [Sarwar et al., 2001], the authors analyze, amongst other parameters, the sensitivity of the parameter k for an item-based algorithm and a regression-based algorithm. Based on their empirical results, Sarwar et al. select 30 as their optimal choice of neighborhood size. Herlocker et al. report similar results and consider a neighborhood size of 20 to 50 as reasonable [Herlocker et al., 2002].

Aggregation of ratings

After the k nearest neighbors of user u are determined, their rating values for the target item i have to be combined together to form the prediction value $\hat{r}_{u,i}$. The standard prediction formula of the early collaborative filtering approach proposed by [Resnick et al., 1994] is defined in Equation (2.5).

$$\hat{r}_{u,i} = \overline{r_u} + \frac{\sum_{n \in N} sim(u,n) * (r_{n,i} - \overline{r_n})}{\sum_{n \in N} sim(u,n)} \tag{2.5}$$

The weighting factor $sim(u,n)$ is a measure of similarity between the target user u and one of his or her neighbors $n \in N$ and can be computed with one of the similarity metrics presented above. Note that various other calculation schemes exist for computing a rating prediction. However, one advantage of using the standard scheme is that the evaluation results are easier to compare [O'Donovan and Smyth, 2005a].

Recent collaborative filtering approaches

Over the last decade, a variety of collaborative filtering algorithms have been proposed in both academia and industry. Boosted by the 1 million dollar Netflix Prize competition[3], in particular in the last few years a variety of sophisticated algorithms have been proposed, which rely, e.g., on matrix factorization, probability theory and advanced machine learning techniques [Lawrence and Urtasun, 2009; Koren, 2010]. In the following, we will summarize a small selection of recent approaches, in particular proposed by the author of this thesis.

In [Lemire and Maclachlan, 2005], Lemire and Maclachlan propose the *Slope One* family of item-based recommender algorithms. Slope One predictors have the form $f(x) = x + b$ and are based on the computation of "popularity differentials between items for users". Their evaluation shows that the Slope One family leads despite its simplicity to relatively accurate predictions. Due to its simplicity, different implementations of the algorithm in various programming languages and frameworks are available today.

One of the main problems of collaborative filtering is the cold start problem [Hu and Pu, 2011; Liu et al., 2011], that is, how to filter items and make recommendations when the user community is small and the data set is sparse [Huang et al., 2004; Ma et al., 2011]. The problem can be divided into the subproblems new user and new item problem [Schein et al., 2002]. The new user problem deals with the question of how to compute recommendations for new users who did not provide any rating information

---

[3]http://netflixprize.com

yet. Analogously, the new item problem deals with the question of how to recommend new items for which no rating information yet exists.

In the line of this research, we recently presented a novel collaborative filtering approach called *RF-Rec* (rating frequency recommender) [Gedikli and Jannach, 2010d,e], which in particular improves the capability to deal with sparse data sets. In contrast to other collaborative filtering approaches the rating prediction for a user u and an item i not yet seen by u is solely based on the information about frequencies of rating values in the database. If, for example, the most frequent rating for item i in the database is 4 and user u's most frequent rating is also 4, the algorithm will basically also predict 4 as u's rating for i.

In [Gedikli et al., 2011a] we propose extensions to the RF-Rec approach in order to further increase the predictive accuracy by introducing schemes to weight and parameterize the components of the predictor. An evaluation on three standard test data sets reveals that the accuracy of our new schemes is higher than traditional collaborative filtering algorithms in particular on sparse data sets and on a par with a recent matrix factorization algorithm. At the same time, the key advantages of the basic scheme such as computational efficiency, scalability, simplicity and the support for incremental updates are still maintained.

In [Gedikli and Jannach, 2010a,b] we propose a collaborative filtering scheme which relies on the data mining technique association rule mining [Agrawal and Srikant, 1994] for generating recommendations. For each user a personalized rule set based on the user's neighborhood is learned using the recent IM-SApriori algorithm [Kiran and Reddy, 2009]. At recommendation time these personalized rule sets are combined with the neighbors' rule sets to generate item proposals. The rule sets contain rules such as *"If Bob likes the movie Heat then he will also like the movie Casino"*, i.e., if *Bob* watched *and* liked the movie *Heat* and did not watch the movie *Casino* before, the movie *Casino* can be recommended to *Bob*. The evaluation of the new method on common collaborative filtering data sets shows that our method outperforms both the IMSApriori recommender as well as a user-based k nearest neighbor method using the Pearson correlation coefficient. The observed improvements in predictive accuracy are particularly strong for sparse data sets.

2.1.3   Content-based recommendations

While collaborative filtering recommender systems recommend items similar users liked in the past, the task of a content-based recommender system is to recommend items that are similar to those the target user liked in the past [Pazzani and Billsus, 2007].

We illustrate the basic rationale of a content-based recommendation method with an example from the movie domain. Figure 2.3 represents an excerpt from an example movie database which also provides plot keywords for each movie[4], i.e., an item's content description is represented by a set of plot keywords. Figure 2.4, on the other hand, shows an excerpt from the user database. The user database maps each user to one or more movies he or she watched in the past. In this example we assume that users only watched movies they like.

Movie Title	Plot Keywords			
Heat (1995)	Detective	**Criminal**	Thief	**Gangster**
Scarface (1983)	**Gangster**	**Criminal**	Drugs	Cocaine
Amélie (2001)	**Love**	Waitress	Garden Gnome	Happiness
Eat Pray Love (2010)	Divorce	India	**Love**	Inner Peace
...	...			

Figure 2.3: Movie data set with content description.

A simple content-based recommender computes recommendations for *Alice* by selecting movies *Alice* is not aware of and which are similar to those movies she watched before. Note that in this simple example similarity between movies could be defined by the number of overlapping keywords. Formally,

---

[4]The plot keywords were taken from the Web site IMDb.com.

User	Preference Profile
Alice	Eat Pray Love (2010), What Women Want (2000)
Bob	Scarface (1983), Carlito's Way (1993), Terminator II (1991)
...	...

Figure 2.4: User data set.

this idea is captured by the Dice coefficient [Baeza-Yates and Ribeiro-Neto, 1999]:

$$sim(a, b) = \frac{2 \times |keywords(a) \cap keywords(b)|}{|keywords(a)| + |keywords(b)|} \tag{2.6}$$

where a and b represent the items to be compared and the function *keywords* returns the corresponding set of keywords of a given item. The possible similarity values are between 0 and 1 inclusive. The unseen movie *Amélie*, for example, has one keyword ("Love") in common with *Eat Pray Love*. The Dice coefficient returns $\frac{2}{8}$ in this case. Therefore, we can assume some degree of similarity between both movies. Since *Alice* liked *Eat Pray Love* in the past, *Amélie* is finally recommended to her.

As described above, a traditional collaborative filtering algorithm takes a user-item rating matrix as the only input, whereas a content-based recommender needs both the preferences of the target user as well as the textual description of the items to be recommended – the "content". Note that for a content-based recommender no user community is required for generating recommendations. Still the cold start problem exists as the target user has to provide an initial list of "like" and "dislike" statements. However, in a content-based recommender new items can be incorporated easily in the recommendation process because similarity to existing items can be computed without the need for any rating data.

Note that in the example discussed above we did not take the importance of each keyword into account, that is, each keyword gets the same importance. However, it appears intuitive that keywords which appear more often in descriptions are less representative. Next we will show how more sophisticated metrics from the field of information retrieval address such problems and how they can be incorporated in a content-based recommender.

Content representation

In collaborative filtering two items are similar when different users have provided similar ratings for these items. Therefore, except the user-provided ratings, no additional information about the items is needed. On the other hand, in a content-based approach similarity is defined with respect to content. However, a uniform content representation of items is necessary in order to measure and compare similarity between items. Several ways exist to represent the content of an item. In the movie domain, for example, a movie's content description can be represented by using a list of features such as director, actors, genre, description, and related-titles. If the items to be recommended are text documents such as Web sites or research articles no additional effort is necessary for acquiring the item content as they are per se descriptive, and the relation to information retrieval systems gets obvious. Therefore, content-based recommenders have their roots in information retrieval [Hanani et al., 2001; Melville and Sindhwani, 2010] where the task is to find items that match the users' information needs and filter out unrelated items.

For convenience, we will assume in the following that the underlying item set consists of text documents. Note that item descriptions which are composed of different slots, such as director and actors in the movie domain, can be viewed as single text documents. One intuitive way to represent the (textual) content of items is to use a binary representation of the content, that is, a binary vector which indicates whether pre-selected words or phrases appear in the text document or not. However, this naive approach does not take the importance of each word and the length of each document into account. Due to the shortcomings of the naive approach, the TF-IDF encoding format was proposed and gained popularity in the field of information retrieval [Salton et al., 1975; Baeza-Yates and Ribeiro-Neto, 1999]. TF-IDF stands for term frequency - inverse document frequency and is used to determine the relevance of terms in documents of a document collection. As the name suggests the TF-IDF measure is composed by two frequency measures.

The idea of the term frequency measure $TF(i,j)$ is to estimate the importance of a term i in a given document j by counting the number of times a given term i appears in document j. Additionally, a normalization is possible, e.g., by dividing the absolute number of occurrences of term i in document j by the absolute number of occurrences of the most frequent word in document j. Several other schemes are possible.

On the other hand, the idea of the inverse document frequency measure $IDF(i)$ is to capture the importance of a term i in the whole set of available documents. Therefore, $IDF(i)$ can be seen as a global measure which reduces the weight of words that appear in many documents, e.g., stop-words such as "a", "by", or "about", since they are usually not representative and helpful to differentiate between documents. Formally, inverse document frequency is usually computed as

$$IDF(i) = log\frac{N}{n(i)} \tag{2.7}$$

where N is the size of the document set and $n(i)$ is the number documents in which the given term i appears. We assume that each term appears in at least one document, i.e., $n(i) \geq 1$. If $n(i) = N$ the logarithm function returns 0 indicating that term i is of no importance for discriminating between documents as it appears in all documents.

Finally, the TF-IDF measure, which represents the weight for a term i in document j, is defined as the combination of these two measures:

$$\text{TF-IDF}(i,j) = TF(i,j) * IDF(i) \tag{2.8}$$

With the help of the TF-IDF measure, text documents, or generally speaking the textual description of items, can be encoded as a TF-IDF weight vector. Note that many improvements have been made to the basic TF-IDF vector space model such as reducing the number of dimensions by performing stemming [Porter, 1997] and removing irrelevant stop words or uninformative keywords [Pazzani and Billsus, 2007]. Next we will see a content-based recommendation technique that relies on the vector space model.

Recommending similar items

The k nearest neighbor method can also be incorporated in a content-based filtering recommender to compute recommendations. For predicting a rating value $\hat{r}_{u,i}$ the k most similar items to the target item i, for which target user u provided a rating, are determined using the items' content description represented by a TF-IDF vector. Afterwards, the user's ratings for the most similar items can be aggregated to form the prediction value. A nearest neighbor content-based filtering approach is, for example, applied in [Billsus et al., 2000]. The authors present a content-based learning agent for wireless access to news which can lead to a reduction of bandwidth, time, and transmission costs. Preference information is collected implicitly by observing a user's actions such as selecting or skipping a news story. The learned user models are used to compute a personalized list of news items.

Recent content-based approaches / hybrids

Pure content-based filtering methods typically match the user profiles with the content representation of items ignoring data from other users. In [Mooney and Roy, 2000], for example, Mooney and Roy present a content-based book recommender system. Each book is represented by a list of slots such as title, authors, synopses, published reviews, and customer comments. The information for each slot was collected from the Web pages at Amazon.com. Mooney and Roy use a Bayesian learning algorithm to compute a ranked list of book titles. While the authors show that their approach can produce accurate recommendations, in practice, however, hybrid approaches are often used which combine both techniques a content-based filtering method and a collaborative filtering approach [Balabanovic and Shoham, 1997a; Good et al., 1999; Soboroff and Nicholas, 1999; Popescul et al., 2001; Li and Kim, 2003]. Such a hybrid approach may help overcome the limitations of both approaches. A pure collaborative filtering approach is not capable of providing recommendations in cold-start situations [Schein et al., 2002], while a pure content-based approach is usually not as accurate as a collaborative filtering approach [Pilászy and Tikk, 2009]. A hybrid approach tries to combine the advantages of both methods [Balabanovic and Shoham,

1997b; Melville et al., 2002; Adomavicius and Tuzhilin, 2005]. In [Luo et al., 2009], for instance, the authors present a framework for personalized news video recommendation where different information sources are exploited for selecting news topics of interest to the user. Their proposed system supports an interactive topic network navigation and exploration process, and additionally it can recommend news videos of interest from a large-scale collection of news videos. The algorithm for ranking the news videos according to their importance and representativeness takes into account the news topic of a particular news video v, the broadcast time of v on a particular TV news program, the visiting times for v of all users, the rating score of v of all users, and the video quality in terms of frame resolution and video length. Their experimental evaluation of the algorithm on a collection of news videos[5] reveals that the algorithm yields high-accuracy results.

2.2 Evaluating recommender systems

In this section we want to review how the quality of a recommender system with respect to different aspects can be assessed. In particular we want to focus on the different types of evaluation that exist in the recommender system literature. In [Shani and Gunawardana, 2011], Shani and Gunawardana identify three types of experiment:

- offline experiments (Section 2.2.1),
- user studies (Section 2.2.2), and
- online experiments (Section 2.2.3).

The typical characteristics of the different evaluation types are illustrated in Figure 2.5.

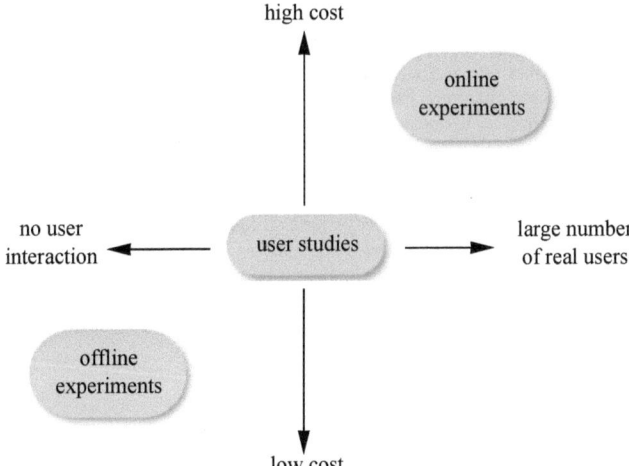

Figure 2.5: Typical characteristics of different evaluation types.

It can be seen from Figure 2.5 that offline experiments do not require user interaction at all. Therefore, they can be conducted offline at low cost and are commonly used in the literature. User studies refer to an experiment setting where a small group of users interact with the system and report their experience with the system, typically by answering a set of questions. Online experiments, on the other hand, represent large scale experiments with real users of a running system over a longer period of time. User studies and online experiments are more costly than simple offline experiments and are harder to conduct. In the following, we will explain the different evaluation types and their application areas in more detail.

---

[5]The data set consists of news videos captured from 3 TV news channels for more than 3 months.

2.2.1 Offline experiments

Offline experiments are easy to conduct and widely used in the literature because no interactions with real system users are required. Usually historical user transaction data is used in offline experiment settings. The freely available MovieLens data sets[6], for example, are often used in the recommender system literature [Zhang and Pu, 2007; Yildirim and Krishnamoorthy, 2008; Symeonidis et al., 2009; Gedikli and Jannach, 2010d]. Together with the Netflix data set, which became popular due to the Netflix Prize competition[7], they build a collection of accepted standard data sets which can be used to make the results comparable to existing approaches. Such a data set typically contains user preferences for a given item set. These user preferences are used to simulate the behavior of real users. A simple test protocol would look like follows: The user preferences are randomly assigned to a train or test set. The user preferences in the train set are used to build a user profile which then is used to predict the user's hidden item preferences in the test set. If the task to be evaluated is a prediction task, i.e., the question how accurate can the recommender algorithm make predictions, one can simply check how far the predicted value corresponds to the real preference value. Later on in this section, we will present different accuracy metrics which can be used to assess the prediction quality of a recommender.

If the used data set is publicly available and the evaluation protocol is described in detail, offline experiments can even be repeated by other researchers. In the Netflix Prize competition, for example, both the data set as well as the evaluation protocol were predefined by the organizers for the sake of reproducibility of the results.

The experiment designer has to choose the data carefully because some of the available data sets come with a bias in the data. For example, the 100k-MovieLens data set only contains users who have rated at least 20 items; the minimum number of rated items per user in the Yahoo!Movies data set[8] is 10; the Book-Crossing data set[9] [Ziegler et al., 2005] only contains users with at least one rating. These are examples for biases which can be found in publicly available data sets. The experiment designer should be aware of these data biases and try to correct them, e.g., by sampling the data appropriately.

In [Shani and Gunawardana, 2011], the authors point to another important drawback of offline experiments, that is, the number of questions which can be analyzed through offline experiments is limited. Usually the prediction power of recommender algorithms is analyzed in offline experiments. For more complex question types, however, user studies and online experiments are used, which will be presented in the subsequent sections.

Note that comparative offline experiments are also reported in this thesis. In Chapter 3 we come to the conclusion that our tag recommender algorithm LocalRank outperforms another algorithm because an offline experiment on a popular data set reveals that our algorithm is more efficient and has a slightly better recommendation accuracy than the baseline recommender. In Chapter 4 offline experiments are conducted to evaluate which of the tag-based algorithms performs best.

In the next section, we will describe popular evaluation metrics and procedures commonly found in offline experiment settings.

Evaluation metrics and procedures

The *Mean Absolute Error* (MAE) is a widely accepted evaluation metric in recommender system research. It can be used to measure the predictive accuracy of recommender algorithms. The MAE metric measures the average absolute deviation between the predicted rating by a recommender and a user's real (but withheld) rating. MAE can be formally defined as

$$MAE = \frac{\sum_{i=1}^{N} |\hat{r}_i - r_i|}{N} \tag{2.9}$$

where N is the number of tested user-item combinations. The lower the MAE value is, the better a recommender can predict a person's feelings towards an item.

---

[6] http://www.grouplens.org/node/73
[7] http://netflixprize.com
[8] http://webscope.sandbox.yahoo.com
[9] http://www.informatik.uni-freiburg.de/~cziegler/BX

The *Root Mean Squared Error* (RMSE) is another standard metric in recommender system research to measure the predictive accuracy of recommender algorithms. RMSE is a statistical accuracy metric which gained popularity thanks to the Netflix Prize. Similar to the MAE, the RMSE metric measures the average absolute deviation between the predicted rating by a recommender and a user's real (but withheld) rating, but puts more emphasis on larger prediction errors. Formally,

$$RMSE = \sqrt{\frac{\sum_{i=1}^{N} |\hat{r}_i - r_i|^2}{N}} \qquad (2.10)$$

where N stands for the total number of predictions. Lower RMSE values obviously correspond to more accurate predictions.

Beside accuracy, the *prediction coverage* of the recommenders can also be easily measured in offline experiments. Prediction coverage is defined as the percentage of user-item combinations a recommender can make predictions [Herlocker et al., 2004] and is calculated as follows:

$$COV = \sum_{i=1}^{N} \frac{\mathbf{1}(\hat{r}_i)}{N} \qquad (2.11)$$

The parameter N is again the number of tested user-item combinations, whereas the indicator function returns 1 if a prediction can be computed and 0 otherwise.

We have so far presented metrics which focus on the prediction task of a recommender system alone. As said before another common task of a recommender system is to provide a list of top-n recommendations, that is, the system has to suggest a ranked list of items the active user is not aware of and will like the most. The quality of these recommendations produced by recommender algorithms can be measured with the standard information retrieval metrics *precision* and *recall*, which were originally proposed in [Cleverdon and Kean, 1968].

In order to determine precision and recall in a recommendation scenario, the item set has to be divided into two classes: interesting (relevant) and not interesting (not relevant) to the user. If the rating scale is not binary, the rating scale has to be transformed first. For example, the 100k-MovieLens rating database[10] contains ratings on a scale from 1 to 5. Herlocker et al. describe a simple way of converting every rating of 4 and 5 to "like" statements and all ratings from 1 to 3 to "dislike" statements [Herlocker et al., 2004]. Note that the question of which items are relevant and which are not has a big influence on the computation of the precision and recall values. In [Harter, 1996], the authors point to this problem of "relevance assessment" and its effects on information retrieval metrics. They propose to take the human factor into account instead of using objective relevance measures. For this reason, the rating transformation proposed by [Sandvig et al., 2007] is personalized to the user. Ratings above the user's mean rating are interpreted as "like" statements and otherwise as "dislike" statements. In this way, Sandvig et al. try to simulate a user's personal decision of which items are relevant and which are not.

After the transformation is done, we denote with ELS the set of existing like statements. The set of predicted like statements returned by a recommender shall be denoted as PLS. Precision is the ratio of the number of like statements in the recommendation list to the length of the recommendation list, formally precision can be defined as

$$precision = \frac{|PLS \cap ELS|}{|PLS|} \qquad (2.12)$$

and measures the number of correct predictions in PLS. Precision can also be interpreted as the probability that an item in the recommendation list is of interest to the user [Herlocker et al., 2004].

On the other hand, recall is defined as the ratio of the number of like statements in the recommendation

---

[10]http://www.grouplens.org/node/73

list to the total number of all like statements, formally recall is measured as

$$recall = \frac{|PLS \cap ELS|}{|ELS|} \qquad (2.13)$$

and describes how many of the existing like statements were found by the recommender.

In the evaluation procedure, recommendations and the corresponding precision and recall values are calculated for all users in the data set and then averaged. Note that both precision and recall values have to be considered simultaneously because improving one is usually at the cost of the other. Therefore, the averaged precision and recall values are combined in the F1-score, where

$$F1 = 2 * \frac{precision * recall}{precision + recall} \qquad (2.14)$$

2.2.2   User studies

As opposed to offline experiments, user studies require much more user effort and time [Wildemuth, 2003]. In a user study typically a small group of volunteer test persons are recruited for the experiment. The test persons are expected to interact with a recommender system and to report their experience with the system. The experiment is either conducted in a laboratory environment or in some other locations, e.g., in private locations. The participants are monitored and interviewed either before, during, or after the experiment. Usually, the results from a user study are used to test some hypotheses which have been formulated by the researcher before the experiment was actually designed. When conducting a user study the question for the appropriate number of users to be recruited for the experiment arises. We can consider the number of recruited users for an experiment large enough when the results are statistically significant according to a statistical significance test [Demšar, 2006; Shani and Gunawardana, 2011]. It is important to know that the statistical significance test has to be selected carefully because some of the tests are either not strong enough to detect existing significant differences or may even lead to false detections of significance in the data where there is no significance at all, see, for example, [Demšar, 2006] and [Smucker et al., 2007] for a comparison of statistical significance tests.

User studies are more costly than simple offline experiments and harder to conduct, but amongst the three different evaluation types discussed here, they can perhaps answer the widest range of question types according to [Shani and Gunawardana, 2011]. In a user study both types of results – quantitative and qualitative results – can be collected. Quantitative results can be recorded by monitoring the user behavior. A quantitative result would be, for instance, the time needed by a user to complete a task which can be measured implicitly. In [Gedikli et al., 2011b], for example, we measure how different explanation interfaces can reduce the user's decision-making time (see also Chapters 5 and 6). We use a direct measurement and compute the time difference for completing the same task of decision making with and without an explanation facility or across different explanation facilities.

On the other hand, qualitative data can also be collected in a user study, e.g., by explicitly asking questions to the users related to their feelings towards the recommender system. Qualitative data can, for example, stand for a user's opinion about a generated recommendation. In [Lewis, 1995], questionnaires are proposed which can be used to measure the usability of a recommender system from the user's perspective. Qualitative data is hard to obtain but necessary to assess the real value of a recommender system. For instance, a user may not always be satisfied with highly accurate recommendations, e.g., when recommending the movie "Terminator II" to a user who already watched the first part of the movie. The probability is high that the user is already aware of the recommended movie. Therefore, this recommendation would be highly accurate but not very useful, that is, accuracy does not always correlate with user satisfaction. See, for example, [McNee et al., 2006] or [Cremonesi et al., 2011] for a broader discussion of this problem. User studies have shown to be a helpful tool for interpreting the quantitative results such as a recommendation list returned by a recommender system. Note that user studies represent the only experiment setting where qualitative data can be collected [Shani and Gunawardana, 2011].

According to [Greenwald, 1976] there are principally two design types of user studies: between-subjects and within-subjects experiments. In a user study typically two or more treatments (systems, algorithms, etc.) are compared with each other with regard to the same task. In a *between-subjects* user study each test user is randomly assigned to one treatment, that is, each user is only confronted with one treatment. Note that between-subjects experiments are also referred to as A-B tests in the literature. Examples for a between-subjects experiment setting can be found in [Sen et al., 2007; Jannach and Hegelich, 2009]. On the other hand, in a *within-subjects* user study each test user is confronted with all treatments to be evaluated, typically in random order. A within-subjects study design can be found in [Vig et al., 2009; Baur et al., 2010; Ge et al., 2012]. For a deeper discussion on the right choice of experiment design and the particular advantages and disadvantages of both experiments the reader is referred to [Greenwald, 1976] and [Shani and Gunawardana, 2011].

According to [Shani and Gunawardana, 2011] within-subjects experiments are more informative than between-subjects experiments, but the test users may be aware of the experiment purpose which, on the other hand, can lead to biasing effects. In between-subjects experiment settings, however, the users are usually not aware that they are part of an experiment making this evaluation scenario more realistic.

Recently, several works have been proposed in the literature that systematically build a user-centric evaluation framework for recommender systems which includes a standardized form of a questionnaire [Chen and Pu, 2010; Pu et al., 2012]. In the work of [Chen and Pu, 2010], for example, a user evaluation framework for recommender systems is presented which tries to explore the evaluation issues from the user's perspective. Their proposed framework is based on findings from decision theory and trust issues and includes both an accuracy and effort measurement model and a user-trust model. In particular, Chen and Pu propose a sample questionnaire to measure the subjective variables in the models.

In [Pu et al., 2012], a user-centric measurement framework called *ResQue* (recommender systems' quality of user experience) is presented. ResQue consists of 15 evaluation constructs, ranging from *perceived system qualities* (users' perception of the objective characteristics such as recommendation quality) and *beliefs* (higher level perception of the system, e.g., system effectivity) through to *attitudes* (users' overall feeling towards a recommender) and *behavioral intentions* (the influence of recommendations on users' decision making). A questionnaire with a total of 32 questions is presented with at least one question for the assessment of each construct. Pu et al. validate ResQue and the causal relationships among its constructs by conducting a Web-based survey. The ResQue questionnaire can be adopted in user studies to assess user acceptance and user attitude towards a recommender. For an example application of the ResQue framework the reader is referred to [Hu, 2012].

We conclude this introduction with an example of a user study reported in [Swearingen and Sinha, 2002]. The authors try to analyze evaluation factors different from statistical accuracy metrics such as the MAE by analyzing user interaction with 11 online recommender systems in a within-subjects user study. They address questions of the form *"What factors lead to satisfaction with a recommender system?"* and *"What encourages users to reveal their tastes to online systems, and act upon the recommendations provided by such systems?"*, which are hard to answer in offline experiments. In their experiment setting the recommender systems were presented in random order to the users in order to account for biasing effects. The users were expected to interact with the systems, that is, the users provided user preferences in order to get useful recommendations. They were then asked to evaluate the recommendations of each system on different evaluation dimensions such as liking, transparency, and familiarity, which are hard to measure in offline experiments. Additionally, an overall rating for each recommender was expected from the users. Swearingen and Sinha conclude with a general design guideline for recommender systems developed from the user's perspective. For example, the authors suggest to ask the users kindly to provide a few more ratings if that increases recommendation accuracy because, according to the authors, "users dislike bad recommendations more than they dislike providing a few additional ratings". Their study also stress the important role of transparency (understanding of system logic) in recommender systems.

Comparative user studies on the user's perception of recommender systems have, for example, been conducted by [Felfernig and Gula, 2006; Ricci and Nguyen, 2007; Celma and Herrera, 2008; Krishnan et al., 2008; Baur et al., 2010]. Comparative user studies can also be found in this thesis. In the Chapters 5 and 6, we report the results of two user studies on the user's perception of different explanation interfaces.

2.2.3 Online experiments

Finally, we want to introduce the last category of experiments that can be used to compare the performance of several recommender systems: online experiments [Shani and Gunawardana, 2011]. Compared to the experiments presented so far, online experiments are conducted on a running system with real system users in order to reliably evaluate new ideas. They represent large scale highly interactive experiments, that is, experiments with a large number of users. Typically such online experiments are designed to understand the behavior of users in real-world settings over a longer period of time, e.g., over several months. One can, for example, design an online experiment to understand the real effects of a recommender algorithm such as the increase of the conversion rate which measures the ratio of the number of Web site visitors to the number of buyers for an online retailer [Jannach and Hegelich, 2009]. As already reported in [Jannach and Hegelich, 2009], only a few real-world studies exist because online experiments are more costly than simple offline experiments and are harder to conduct but their results can have a bigger impact. Next we will provide a popular example for an online experiment in a Web environment.

Greg Linden of Amazon writes in his blog about an interesting project he did at Amazon [Linden, 2006a,b]. He created a prototype and modified the Amazon.com shopping cart Web page and provided product recommendations of interest to the user which are related to the items in the shopping cart. However, a marketing senior vice-president was against this idea because he was afraid that it might have a negative impact on sales. Nevertheless, Greg conducted an online experiment in order to measure the sales impact, although he was explicitly forbidden to continue his work on this project. The test results showed clearly that the feature he proposed "won by such a wide margin that not having it live was costing Amazon a noticeable chunk of change". This story is a good example showing the power of such online experiments where real system users are involved.

In [Kohavi et al., 2009], Kohavi et al. provide a practical guide for such online experiments on the Web. In their simplest form, online experiments typically compare one or more different approaches with the existing baseline approach. To do so, system users are randomly assigned to groups, where the users of each group are only confronted with one approach. The user group assigned to the existing approach is often referred to as the control group. Usually, the traffic of a running system is redirected accordingly to different alternative groups [Shani and Gunawardana, 2011]. Note that this experiment setting corresponds to a between-subjects setting (see Section 2.2.2). The different approaches (variants) compared in an online experiment can, for example, represent different algorithms or different kinds of user interfaces, which are tested against each other to find out which performs best. Again, statistical tests are used to decide whether the observed results are significant or not [Demšar, 2006].

Note that in general experimental studies, it is of high importance to keep all untested variables fixed because otherwise the differences in the results cannot be attributed to the tested variable alone. For example, suppose that we want to promote a new user interface for recommender systems. Therefore, we test whether a recommender system A with the newly proposed user interface can outperform an existing one B. We assume that outperform means A is able to increase sales and profit compared to B. If A uses another recommender algorithm to generate user recommendations than B, no reliable conclusions can be drawn from the results. For example, if A actually outperforms B, then the researcher is not able to tell that the differences are due to the newly proposed user interface, as the results can also be explained by the superior performance of the underlying recommender algorithm. Therefore, when designing an experiment care has to be taken to ensure that no bias is introduced in the study. Otherwise, the obtained results are questionable.

The application of online experiments can sometimes be risky as it can cause undesired effects [Shani and Gunawardana, 2011]. The experimenter has to keep in mind that the evaluation is conducted with real system users. Therefore, potential long term issues may arise. For example, when the alternative system presented to the users in an A-B test is poorly engineered, users may stop using the system in general which is unacceptable for many online retailers and which has to be avoided. This is also the reason why the marketing senior vice-president in the Amazon example was "dead set against" the feature proposed by Greg Linden because he was afraid that the proposed feature can negatively impact sales.

For these reasons, [Shani and Gunawardana, 2011] recommend ways to reduce or avoid those risks. They suggest to apply online experiments in the end of the evaluation chain, after "an extensive offline study provides evidence that the candidate approaches are reasonable, and perhaps after a user study that measures the user's attitude towards the system". Although subsequent runs of different experiment types are time-consuming and expensive, they gradually reduce the risk of frustrating the users and make the results more representative. However, note that only a few studies which make use of different experiment types, are available in the literature, see, for example, [Ziegler et al., 2005] or [Gedikli and Jannach, 2013]. In [Gedikli and Jannach, 2013], for example, we compare different explanation styles for recommender systems. First, we run offline experiments in order to learn the optimal parameters for the parameterized explanation interfaces. Afterwards, these interfaces are compared in a within-subjects user study.

In summary, it can be said that online experiments represent powerful means to systematically evaluate proposed hypotheses in order to test new ideas for recommender system design or improvement. Online experiments can be used to directly measure long-term goals of a recommender system such as user loyalty or the impact on sales and profit. However, as described above online experiments entail the risk that the experiment can have a negative effect on the users if the experiment design has not been adequately thought through. In the worst case, the user will stop using the system which is not acceptable in commercial applications. The risks of such experiments can be reduced when the tested alternative systems are technically mature and already passed prior offline experiments and/or user studies.

2.3 Recommendations based on Social Web tagging data

We have so far introduced the basic recommendation techniques collaborative and content-based filtering and discussed three types of experiments which are commonly used for evaluations in the recommender system literature. In the further course of this thesis, we will introduce new recommendation approaches which are based on the data and concepts of the Social Web [Kim et al., 2010b]. In particular, we will exploit Social Web tagging data for recommender systems. Therefore, a selection of recent developments in the field of tag-based recommender systems is presented later on in this section.

In the last years more and more recommender systems were presented which try to outperform the existing ones by exploiting additional, external information. For example, hybrid recommendation approaches try to exploit additional information about items or users and combine different techniques to achieve better accuracy results than pure collaborative and content-based filtering approaches [Adomavicius and Tuzhilin, 2005; Jannach et al., 2010; Melville and Sindhwani, 2010]. Similarly, multi-criteria recommender systems incorporate multi-criteria rating information in the recommendation process to further improve the predictive accuracy [Adomavicius and Kwon, 2007; Jannach et al., 2012].

The Semantic Web opened new doors for exploiting additional data sources for recommender systems. One can, for example, use the additional knowledge encoded in classification taxonomies available on the Web. A plethora of hand-crafted classification taxonomies exist for various domains such as books, CDs, DVDs, and electronic goods. Ziegler et al., for example, propose an algorithm to increase diversity of the items in the recommendation list by exploiting the Amazon.com's book taxonomy [Ziegler et al., 2005]. They propose to compute item similarity based upon a taxonomy-based similarity metric [Ziegler et al., 2004]. In order to show the superiority of their diversifying algorithm, the authors also introduce a new intra-list similarity metric. Interestingly, the results indicate that diversifying recommendation lists of an item-based collaborative filtering algorithm is more effective than diversifying the lists of its user-based counterpart, where effectiveness stands for the user's overall liking of recommendation lists.

In [Heitmann and Hayes, 2010], Linked Data, which represents data used to connect content in the Semantic Web [Berners-Lee, 2006], is utilized for recommender systems. Heitmann and Hayes propose to augment the available data for a "closed" recommender system with Linked Data in order to build an "open" recommender that can exploit external data from the Semantic Web. The authors demonstrate for a collaborative filtering music recommender that utilizing Linked Data as additional data source can improve its accuracy in terms of precision and recall.

The advent of the Social Web opened new ways of promoting and sharing user-generated content [Kim et al., 2010b]. Web site visitors turned from passive recipients of information into active and engaged contributors. The Social Web allows users to create and share a large amount of different types of content such as pictures, videos, bookmarks, blogs, comments, or tagging data. It allows users to collaborate with other users on new types of Web applications called Social Web platforms such as Delicious[11], Flickr[12], Facebook[13], and YouTube[14].

Leveraging useful data from the large amount of user-contributed data available in the Social Web represents a challenging topic which opens new opportunities for recommender system research [Guy et al., 2010a; Freyne et al., 2011]. For example, one can think of a recommender that develops and maintains a user profile by analyzing the blog entries the user has commented on recently [Turdakov, 2007]. Such an approach can help to alleviate the cold-start problem of recommender systems. Or think, for example, of a recommender that analyzes the relationships in a social network such as Facebook to compute a user's trust in other users. In the context of recommender systems, trust can serve as a new measure of user-similarity [Golbeck, 2009]. Usually, social networks analysis (SNA) methods are used to analyze the huge amount of data available in a social network [Scott, 2000]. SNA methods can, for example, be used to analyze the role of a particular user in a social network and to determine the existing user clusters. Different works exist in the literature that try to explore SNA methods for recommender systems (see, for example, [Ting et al., 2012] and [He and Chu, 2010]).

Tags are today a popular means for users to organize and retrieve items of interest in the Social Web [Vig et al., 2009]. As the application areas of tags are manifold, they play an increasingly important role in the Social Web. They can be used to categorize items, express preferences about items, retrieve items of interest, and so on. For a detailed description of the main purposes of tagging the reader is referred to a textbook specifically focusing on tagging systems [Peters and Becker, 2009].

Collaborative tagging or social tagging describes the practice of collaboratively annotating items with freely chosen tags [Golder and Huberman, 2006] which plays an important role in sharing content in the Social Web [Ji et al., 2007]. In a social tagging system such as Delicious and Flickr users typically create new content (items), assign tags to these items, and share them with other users [Cantador et al., 2010]. The result of social tagging is a complex network of interrelated users, items, and tags often referred to as a community-created *folksonomy* [Mathes, 2004]. The term folksonomy is a neologism introduced by the information architect Thomas Vander Wal and is composed of the terms *folk* as in people and *taxonomy* which stands for the practice and science of classification [Wal, 2007]. In contrast to typical taxonomies such as formal Semantic Web ontologies, social tagging represents a more light-weight approach, which does not rely on a pre-defined set of concepts and terms that can be used for annotation. For a detailed comparison of both approaches the reader is referred to [Shirky, 2005] and [Ji et al., 2007]. A formal definition of a folksonomy is provided in Chapter 3.

Tags also began to gain importance in the field of recommender systems. User-generated tags not only convey additional information about the items, they also tell something about the user. For example, if two users use the same set of tags to describe an item, we can assume a certain degree of similarity between those. Therefore, tagging data can be used to augment the basic user-item rating matrix.

Recently, several works have been proposed in the literature concerning the topic of leveraging user contributed data available in the Social Web for recommender systems [Guy et al., 2010a; Freyne et al., 2011]. In this work, we pursue this line of research and present novel recommendation approaches. In particular, we will show how tagging data can be utilized for recommender systems. In the following, we will present a possible categorization of tag-based recommender systems in the literature.

2.3.1 Using tags as content

Maybe the easiest way to use tagging data for recommender systems is to consider tagging data as an additional source of content. Several works exist that view tags as content descriptors for content-based systems, see, for example, in [Firan et al., 2007; Li et al., 2008] or [Vatturi et al., 2008]. In Section 2.1.3, we have seen that an item's content description can be represented by a set of keywords. Since tags can be seen as user-provided keywords, the content-based recommendation scheme described in Section 2.1.3 can be applied without any modification.

Similarly, in [de Gemmis et al., 2008], tagging data is used for an existing content-based recommender system in order to increase the overall predictive accuracy of the system. Machine learning techniques

---

[11]http://www.delicious.com
[12]http://www.flickr.com
[13]http://www.facebook.com
[14]http://www.youtube.com

are applied both on the textual descriptions of items (static data) and on the tagging data (dynamic data) to build user profiles and learn user interests. The user profile consists of three parts: the static content, the user's personal tags, and the social tags which build the collaborative part of the user profile. Thus, in this work, tags are seen as an additional source of information used for learning the profile of a particular user. The authors compare their tag-based approach with a pure content-based recommender in a user study. The results show that the recommendations made by the tag-augmented recommender are slightly more accurate than the recommendations of the pure content-based one.

In the study of [Firan et al., 2007] tags are also seen as content descriptors for different content-based systems. Tags are used for building user profiles for the popular music community site Last.fm. In order to address the cold start problem, the user profiles are inferred automatically, e.g., from the music tracks available on the computer of each user, thus reducing the manual effort from the user's side to express his or her preferences. The authors show that tag-based profiles can lead to better music recommendations than conventional user profiles based on song and track usage.

In [Cantador et al., 2010] tags are considered as content features that describe both user and item profiles. Cantador et al. propose weighting functions which assess the importance of a particular tag for a given user or item, and similarity functions which compute the similarity between a user profile and an item profile. These weighting and similarity functions are then combined in different content-based recommendation models. User interests and item characteristics are modeled as vectors $u_m = (u_{m,1}, ..., u_{m,L})$ and $i_n = (i_{n,1}, ..., i_{n,L})$ of length L respectively, where L is the number of tags in the folksonomy, $u_{m,l}$ is the number of times user u_m has annotated items with tag t_l, and $i_{n,l}$ is the number of times item i_n has been annotated with tag t_l. After modeling users and items as vectors accordingly, the authors can adapt the well-known TF-IDF vector space model from information retrieval which was described in Section 2.1.3. Besides this TF-IDF-based profile model, the authors also include a pure TF-based profile model (without the IDF component) into the evaluation pool. Additionally, they propose a profile model based on the Okapi BM25 weighting scheme which is a probabilistic framework to rank documents according to a given query [Baeza-Yates and Ribeiro-Neto, 1999]. These profile models are then exploited in a number of content-based recommendation approaches such as a TF-IDF cosine-based recommendation approach which computes the similarity between a user and an item vector with the cosine similarity measure, and a corresponding BM25 cosine-based recommendation approach. The evaluation results on the Delicious and Last.fm data sets show that the recommendation models focusing on user profiles outperform the models focusing on item profiles.

Tagging data can also be incorporated in search engines to personalize the search results. According to [Pitkow et al., 2002], two basic approaches to Web search personalization can be differentiated. In the first approach, a user's original query is modified and adapted to the needs of the user. For example, the query "eclipse" might be extended to "eclipse software development environment" if we know that the user has an interest in software development. In the second approach, the query is not modified, but the returned list of search results is re-ranked according to the user profile.

An example for the latter approach is given by [Noll and Meinel, 2007]. The authors propose a pure tag-based personalization method to re-rank the Web search results which is independent from the underlying search engine. The basic idea is to use bookmarks and tagging data to re-rank the documents in the search result list. Noll and Meinel also propose a concept called *tagmarking* which translates the keywords in the search query to tags and assign them to the bookmarked Web page that is associated with the query. Bookmarks and tags are aggregated in a binary tag-document matrix M_d where each column (vector) represents a bookmark of a document with its components set to 1 if the corresponding tag is associated with the document and 0 otherwise. The user profile p_u is modeled as a vector $M_d \cdot \omega_d$ where ω_d is a vector which contains the weights assigned to each tag. The tag-user matrix M_u and the document profile p_d are built analogously. Note that by defining $\omega_d := 1^T$ and $\omega_u := 1^T$, the authors assign equal importance to all tags and users. Finally, in the personalization step the documents are re-ranked according to a similarity metric which combines both the user profile and the document profile. Table 2.1 shows in an example from [Noll and Meinel, 2007], how personalization affects Google's result list for the search query "security". The ranking of the Web site of the US Social Security Administration (`ssa.gov`), for instance, has increased because according to the authors the user who submitted the query also shows interest in insurance matters. In the evaluation phase the participants were asked in a questionnaire which of the ranking lists of a query (the original list or the personalized list) they prefer.

The results show that the personalized list was preferred over the original list or, at least, was considered as good as the original list.

Rank	Δ Rank	URL
1	•	securityfocus.com
2	↑ +7	cert.org
3	•	microsoft.com/technet/security/def...
4	↑ +4	w3.org/Security
5	↑ +2	ssa.gov
6	↑ +4	nsa.gov
7	↓ −5	microsoft.com/security
8	↓ −2	windowsitpro.com/WindowsSecurity
9	↓ −4	whitehouse.gov/homeland
10	↓ −6	dhs.gov

Table 2.1: Re-ranking Google's result list for the keyword "security" [Noll and Meinel, 2007].

2.3.2 Clustering approaches

Many tag-based clustering approaches have been proposed in the literature which cluster users and items according to topics of interest by exploiting additional tagging data, see, for example, in [Li et al., 2008; Xu et al., 2011b] or [Zanardi and Capra, 2011].

Li et al. propose in [Li et al., 2008] a system called Internet Social Interest Discovery (ISID) and show its application for the social bookmarking system Delicious[15]. The ISID system, as the name suggests, is a system specifically designed to reveal common user interests based on user-provided tags. The basic assumption, which is then justified in the work, is that user-provided tags are more effective at reflecting the users' understanding of the content than the most-informative keywords extracted from the corpus of a Web page. Therefore, tags are seen as good candidates for capturing user interests. The underlying rationale of the ISID system is to discover tags which are commonly used together. Each of these frequent tag sets define a different *topic of user interests*. Association rule mining [Agrawal and Srikant, 1994] is used to discover the frequent tag sets. Note that this approach is similar to the clustering algorithm proposed in [Brooks and Montanez, 2006] in which tags are used for classifying blog entries. However, the difference is that the blog entries are clustered based on the co-occurrence of a single tag, instead of multiple tags as in the ISID system.

The software architecture of the ISID system is visualized in Figure 2.6. The ISID system gets a stream of posts $p = (user, URL, tags)$ as input where each post p is a combination of a $user$, a URL, and the tags assigned by the $user$ to the URL. The topic discovery component takes these posts as input and returns the frequent tag sets – the topics of interests – as output. Note that beside a list of posts no other information is needed. In particular, no information about offline social connections among users or online connections in a social graph such as Facebook is required. In the clustering component, on the other hand, for each topic of interest (tag set) a user cluster and a URL cluster are identified. The function of the ISID indexing component is to provide different query services for applications which are executed on the discovered clusters and topics of interests. An example query might be *"For a given topic, list all users that are interested in this topic, i.e., have used all tags of the topic"* [Li et al., 2008].

For evaluation purposes, Li et al. tested whether ISID covers the most popular tags of each user. The results show that the ISID topics cover more than 80% tags of 90% of all the users. Moreover, the authors conducted a user study where four reviewers have rated the matching of URL contents with the topics. The resulting assessment of the survey was good to very good.

Similarly, in [Xu et al., 2011b], co-occurring tags are used to build topics of interests. In the resource-tag matrix, each tag is described by a set of resources, to which this tag has been assigned. Afterwards, the authors obtain the tag similarity matrix by computing the cosine similarity between the tag vectors in the resource-tag matrix. Based on this similarity matrix, a graph is constructed where the tags represent the nodes and the edges represent the similarity relationships between the tags. Afterwards, a clustering

---

[15]The data used for evaluation is a partial dump of the Delicious database (http://www.delicious.com).

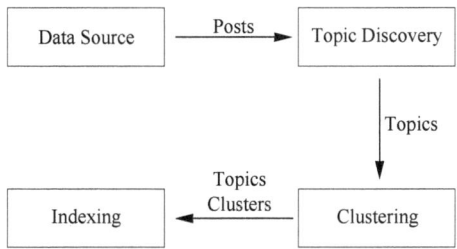

Figure 2.6: ISID system software architecture [Li et al., 2008].

algorithm is used to cluster the tags and extract the topics of interests. Finally, the authors present a topic-oriented tag-based recommendation system (TOAST). TOAST applies preference propagation on an undirected graph called topic-oriented graph which consists of three kinds of nodes: users, resources, and topics. In their recommendation strategy the authors then propagate a user's preference through transitional nodes such as users, resources, and topics, to reach an unknown resource node along the shortest connecting path.

Tag co-occurrence metrics also play an important role in [Sigurbjörnsson and van Zwol, 2008] where tag recommendation strategies for the online photo service Flickr are presented. The authors base their tag recommendation strategies on tags co-occurring with the user-defined tags assigned to a specific photo. In particular, the authors first try to find answers to the questions *"How do users tag?"* and *"What are they tagging?"* and exploit this knowledge in their tag recommendation strategies based on tag co-occurrence.

In [Shepitsen et al., 2008], Shepitsen et al. focus on the recommendation scenario that a user selects a tag and expects a recommendation of resources. They present a recommendation approach which recommends items for a given user-tag pair (u, t). Tag clusters are presumed to act as a bridge between users and items. The idea behind tag clusters is to account for the effects of unsupervised tagging such as redundancy and ambiguity. The authors first determine the items which have some similarity to the query tag t. These items are then re-ranked according to the user profile. The ranking algorithm first calculates the user's interest in each tag cluster as well as the nearest clusters of each item. The nearest clusters are determined by the number of times the item was annotated with a tag from the cluster over the total number of times the item was annotated. Both measures are then combined in the final personalized rank score used to re-rank the item sets. The results show that data sparsity has a big influence on the quality of the clusters which, on the other hand, corresponds with the recommendation accuracy.

In [Zanardi and Capra, 2011], Zanardi and Capra make a distinction between users who tag more items than the average user (the *leaders*), and the others who mainly follow the leaders (the *followers*), e.g., users who mainly browse the content created by the leaders. The leaders are then clustered into domains of interest. Because each user can be interested in multiple topics, a fuzzy clustering method is used to determine the degree a leader belongs to each cluster. The user communities that can best answer a user query, which can be a search or a recommendation, are identified by their proposed clustering approach and used as input for their previously developed algorithm SocialRank [Zanardi and Capra, 2008] to compute an answer. SocialRank first expands the original query such that also similar (related) tags are included in the query to improve coverage. To improve accuracy, on the other hand, the similarity of the taggers is used to rank the returned item set. Therefore, SocialRank computes recommendations in a content-based collaborative-filtering fashion. The evaluation shows that their newly proposed clustering approach achieves comparable accuracy results with the baseline approaches SocialRank and FolkRank[16] [Hotho et al., 2006] in particular on sparse data sets, but has much lower computational costs as it only focuses on the leaders and does not take the whole user database into account.

---

[16]A detailed description of FolkRank is provided in Chapter 3.

2.3.3 Hybrid approaches

Hybrid approaches combine different sources of information to make recommendations. In general, social data such as tagging data is mixed with other types of information such as content data [Seth and Zhang, 2008] or data from the Semantic Web [Durao and Dolog, 2010].

In [Seth and Zhang, 2008], a Bayesian model-based recommender that leverages content and social data is presented. In [Durao and Dolog, 2010], on the other hand, a tag-based recommender which recommends Web pages is extended such that also semantic similarities between tags are discovered which are basically ignored in syntax-based similarity approaches. Table 2.2 shows a motivating example for exploiting semantic relations between tags.

Web page	Tags
$P1$	Programming, Web 2.0, Framework
$P2$	PHP, Scripting, Web 2.0
$P3$	C++, Programming, Framework

Table 2.2: Motivating scenario for exploiting semantic relations between tags.

If we assume a syntax-based similarity measure, the Web pages $P1$ and $P3$ will be considered more similar than $P1$ and $P2$ as $P1$ and $P3$ have two tags in common ("Programming" and "Framework"), whereas $P1$ and $P2$ only share one tag ("Web 2.0"). However, if we analyze the tags in more detail, we see that $P1$ is closer to $P2$ than to $P3$ because $P1$ and $P2$ talk about Web technologies, whereas $P3$ focuses on C++ which is a programming language that is usually not associated with Web technologies. In a semantic-based similarity approach which takes the lexical and social factors of tags into account, these semantic relations can be made explicit. For example, "Web 2.0" would be considered together with "Scripting", and "Programming" with "PHP". Durao and Dolog try to overcome this problem of ignoring the semantic term relations by hybridizing syntax-based approaches such as tag popularity with a new semantic-based approach [Durao and Dolog, 2010]. The authors make use of external semantic sources such as the WordNet dictionary [Voorhees, 1993] and different ontologies from Open Linked Data available on the Web [Bizer et al., 2008] to identify semantic relations between tags which are then considered in the similarity calculations. Their experimental results show an increase of precision if semantic relations are exploited as additional knowledge sources.

A similar hybridization strategy is presented in the work of [Xu et al., 2011a]. In their so called semantic enhancement recommendation framework (SemRec), the idea is to exploit structural as well as semantic information about tags in a unified fusion model. The framework of SemRec is depicted in Figure 2.7.

Hierarchical agglomerative clustering is applied for clustering tags to tackle tag redundancy and ambiguity and to uncover explicit topic models. On the other hand, the authors use Latent Dirichlet Allocation (LDA) analysis [Blei et al., 2003] to reveal implicit semantic relationships hidden in tagging data. Finally, these explicit and hidden topics are combined in a fusion scheme which computes a final score using a parameter λ to weight both components. The fusion model computes the similarity between the target user $\vec{u}$ and each candidate item $\vec{r}$ and is defined as follows:

$$Sim(\vec{u}, \vec{r}) = \lambda \cdot CosSim_{HT}(\vec{u}, \vec{r}) + (1 - \lambda) \cdot CosSim_{TC}(\vec{u}, \vec{r}) \qquad (2.15)$$

The cosine similarities $CosSim_{HT}$ and $CosSim_{TC}$ compute the similarity of the (transformed) vectors over the hidden topics and tag clusters respectively. The items with the top-n similarity scores are then recommended to the target user. The evaluation results on two data sets demonstrate the superiority of the SemRec approach against the pure tag-based recommendation approach of [Noll and Meinel, 2007] and the clustering approach of [Shepitsen et al., 2008] (both approaches are described above).

Tag similarity measures are the focus of the study in [Cattuto et al., 2008]. The authors evaluate five different tag similarity measures with well-established measures of semantic distance. The idea is to map pairs of tags regarded to be similar to a thesaurus such as the WordNet dictionary [Fellbaum, 1998] and to compute the relatedness there with well-defined metrics of semantic similarity. In particular, tags are mapped to so called *synsets* of WordNet which are basically sets of synonyms that stand for one concept. Semantic similarity between two synsets in WordNet is computed in two ways: by measuring the

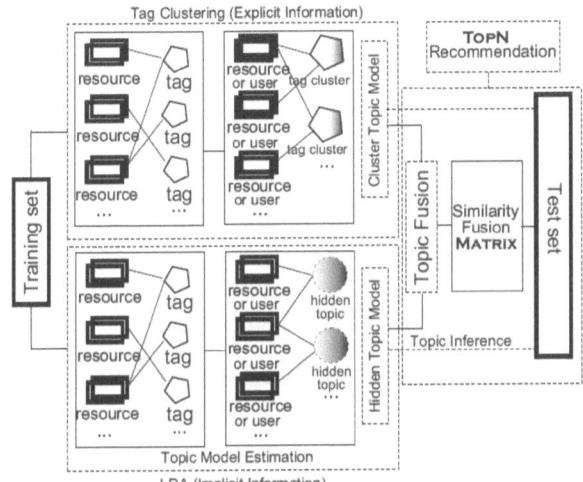

Figure 2.7: SemRec recommendation framework [Xu et al., 2011a].

taxonomic shortest path length and by using the Jiang-Conrath distance measure [Jiang and Conrath, 1997]. Note that according to [Budanitsky and Hirst, 2006] the Jiang-Conrath measure rather represents the human perceived distance between two synsets. The evaluated tag similarity measures are:

1. *Co-Occurrence:* Tag co-occurrence between two tags stands for the number of posts[17] that contain both tags. The most similar tag to the original one is therefore that tag with the highest tag co-occurrence value.

2. *Tag Context Similarity:* This measure computes the cosine similarity between two tag vectors in the vector space $\mathbb{R}^{Tags}$, where for each tag vector $v_t \in \mathbb{R}^{Tags}$ the entries are defined by the tag co-occurrence values. The higher the cosine similarity, the higher the tag similarity (relatedness) is assumed.

3. *Resource Context Similarity:* This similarity is computed analogously, but in the vector space $\mathbb{R}^{Items}$. For each tag vector $v_t \in \mathbb{R}^{Items}$ the entries are defined by the number of times the tag was used to annotate a particular item.

4. *User Context Similarity:* This similarity is computed analogously, but in the vector space $\mathbb{R}^{Users}$. For each tag vector $v_t \in \mathbb{R}^{Users}$ the entries are defined by the number of times the tag was used by a particular user to annotate arbitrary items.

5. *FolkRank:* For each tag t the tags with the highest FolkRank [Hotho et al., 2006] are returned.

The evaluation protocol was designed as follows: For each of the 10,000 most frequent tags of the Delicious data set the most closely related tags were computed with all of the measures presented above. Tags which were not represented in WordNet were skipped. Figure 2.8 shows the average semantic distance from the original tag to the most closely related one for both semantic similarity measures described above. Note that the *random* measure represents a baseline metric which associates each tag with a randomly chosen one.

The results show that the best performance is achieved by the resource context similarity measure. However, the computationally lighter tag context similarity measure achieves similar results. Cattuto et

---

[17] A post is a *(user, item, tags)* tuple which represents the set of *tags* provided by a *user* for a particular *item*.

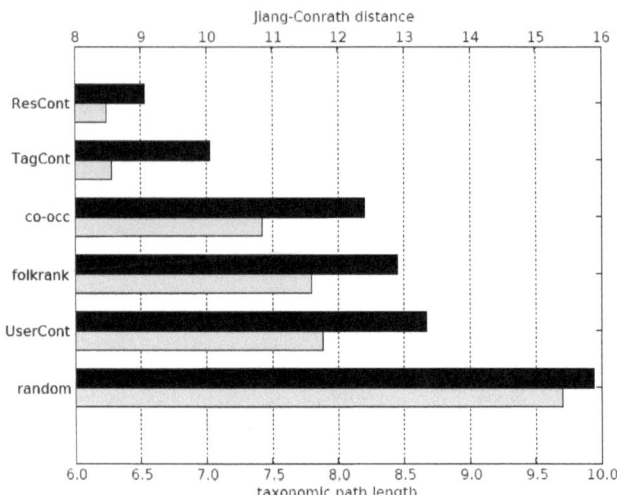

Figure 2.8: Average semantic distance from the original tag to the most closely related one [Cattuto et al., 2008].

al. find out that the tags obtained via tag context or resource context are synonyms or siblings of the original tag, while FolkRank and the tag co-occurrence measure rather provide more general tags.

In [Passant, 2007], Alexandre Passant addresses the limitations of free tagging systems and uses Semantic Web technologies to solve some of these problems. In particular the author identifies the following limitations of free tagging systems:

- *Tags variation:* If syntactically different tags describe the same concept, it is problematic to create the semantic connection between these tags. For example, it is difficult for a tagging system to discover the connection between the tags "high definition" and "HD". Variations can also be caused by simple typo errors.

- *Tags ambiguity:* If one tag can describe different concepts, the system will again not be able to make any difference. The tag "apple", for example, can stand for the fruit or the company with the same name.

- *Flat organization of the tags:* In contrast to ontologies, for example, tags do not form any hierarchy. Again, the semantic relations between tags have to be discovered first, e.g., by using data mining algorithms.

In order to tackle these problems Passant proposes to mix Social Web folksonomies and Semantic Web ontologies. The idea is to link tags to ontology concepts to enhance an information retrieval engine for blog-posts.

In [Mika, 2007], a unified model is presented which covers both social networks as well as semantics. The idea is to extract lightweight ontologies from a folksonomy to better model the concepts of a particular community. For this reason, the traditional bipartite model of an ontology is extended by the user dimension, leading to a tripartite model of actors (users), concepts (tags), and instances (items) which basically corresponds to the tripartite model of a folksonomy. The author shows how two lightweight ontologies based on overlapping communities (O_{ac}) and overlapping sets of items (O_{ci}) can be extracted from the unified model. The network of associations O_{ac}, for example, is built by only considering the associations between actors and concepts. The O_{ci} network, on the other hand, focuses on the associations between concepts and instances.

In the evaluation phase a questionnaire-based evaluation was conducted in which the participants were asked to decide which of the two ontologies (O_{ac} or O_{ci}) is more accurate in terms of the associations between the concepts. The results show that the associations in the community-based network O_{ac} are considered to be more accurate representation of associations between the concepts compared to the associations in the O_{ci} network.

In [Guy et al., 2010b], Guy et al. evaluate people-based and tag-based item recommenders as well as two hybridization strategies for a social media platform. The platform includes different social media applications such as blogs, bookmarks, and wikis. A social aggregation system called SaND [Ronen et al., 2009] is used to aggregate the relationships between users, items, and tags across the different social media applications available on the platform. It basically returns two weighted lists of related users and tags for each user. Note that the SaND system also makes use of information about tags which are applied to users by other users within a people-tagging feature of the platform. The authors propose different types of recommenders which exploit the information provided by the SaND system which are: a people-based recommender (PBR), a tag-based recommender (TBR), two types of a hybrid recommender (PTBR), and a popularity-based baseline recommender (POPBR). The results of a user study show that users are significantly more interested in TBR recommendations than PBR recommendations which again is an indicator for the strength of the user-tag relationships. Note that the differences between TBR and the hybridization strategies, which are viewed by the authors as a variation of a hybrid collaborative filtering and content-based recommender, are not significant. However, hybridization pays off when explanations are included (as discussed later in Section 2.3.5).

The goal of the study in [Bellogín et al., 2010] is to identify the most valuable information sources for recommender systems. The authors analyze the influence of each information source such as ratings, tags, and social contacts on the quality of the recommendations separately and investigate whether and how fusion of these information sources can be beneficial. Recommendation quality is measured on various metrics such as accuracy and diversity. The evaluation results of experiments conducted on a data set obtained from Last.fm show that exploiting tagging data and information about social contacts leads to effective and heterogeneous recommendations.

2.3.4 Tag-enhanced recommenders

There exists a plethora of literature on tag-enhanced recommender algorithms where tagging data is used for improving the performance of traditional collaborative filtering recommender systems. Tagging data is incorporated into existing collaborative filtering algorithms in different ways in order to enhance the quality of recommendations, see, for example, [Ji et al., 2007; Tso-Sutter et al., 2008; Zhao et al., 2008; Liang et al., 2009a; Zhen et al., 2009; Durao and Dolog, 2009; Zhang et al., 2009b; Yuan et al., 2009; Kim et al., 2010a] or [Wang et al., 2010].

In [Tso-Sutter et al., 2008], for example, Tso-Sutter et al. incorporate tags into standard collaborative filtering algorithms. The idea is to reduce the three-dimensional relation $\langle user, item, tag \rangle$ to three two-dimensional relations which are $\langle user, tag \rangle$, $\langle item, tag \rangle$, and $\langle user, item \rangle$, correspondingly. The projection is based on viewing the tags as items ("user tags") and users ("item tags") respectively. For example, in the $\langle user, tag \rangle$ relation tags are viewed as items in the user-item rating matrix. These so called user tags represent tags that are used by the users to tag items. On the other hand, item tags in the $\langle item, tag \rangle$ relation correspond to tags that describe the items. Considering the ternary relation as three two-dimensional relations enables the authors to apply standard collaborative filtering techniques. Tso-Sutter et al. also propose a fusion method which re-combines the individual relations. The results of their empirical analysis show that the predictive performance of their proposed fusion method which incorporates tags outperforms the standard tag-unaware collaborative filtering algorithms.

Exploiting tagging data without reducing the three-dimensional $\langle user, item, tag \rangle$ relation was the next logical step in the literature. In recent years, recommendation methods based on Tensor Factorization (TF) [Tucker, 1966] were proposed which can directly exploit the ternary relationship in tagging data [Symeonidis et al., 2008; Rendle et al., 2009; Rendle and Schmidt-Thie, 2010]. In [Rendle et al., 2009], for example, the authors view the ternary relationship as a three dimensional tensor (cube) and apply the idea of computing low rank approximations for tensors on a tag recommender algorithm. The evaluation results show that their TF-based method achieves even better accuracy results than tag recommender algorithms like FolkRank [Hotho et al., 2006] and PageRank [Brin and Page, 1998]. However, the TF-based model comes with the problem of a cubic runtime in the factorization dimension for prediction and learning. This problem is addressed in the work of [Rendle and Schmidt-Thie, 2010]. Rendle and Schmidt-Thie present a Pairwise Interaction Tensor Factorization (PITF) model with a linear runtime in the factorization dimension. The PITF model explicitly models the pairwise interactions between users, items, and tags.

In [Liang et al., 2008], [Liang et al., 2009a], and [Liang et al., 2009b] extended standard collaborative filtering approaches are presented. In [Liang et al., 2009b], for example, the authors present a tag-based similarity measure to improve standard collaborative filtering approaches. The idea is to cluster users with similar tagging behavior instead of similar rating behavior. In [Liang et al., 2009a], another tag-based method is presented to accurately determine the nearest neighbors of the target user. Liang et al. address the limited tag quality problem [Sen et al., 2007; Bischoff et al., 2008] by building user profiles based on popular tags since according to the authors tag quality is related to tag popularity. Therefore, the authors suggest to map the individual tags of each user and item to these commonly used tags to build a better understanding of the users and the items. Popular tags are referred to as "tags that are used by at least θ users, where θ is a threshold". An experimental evaluation was conducted using a book data set crawled from Amazon.com. The precision and recall results of the compared approaches show that their popularity-based approach outperforms both Tso-Sutter's approach [Tso-Sutter et al., 2008] and Liang's approach [Liang et al., 2008]. The results also indicate that their approach performs better than a recommendation approach based on Singular Value Decomposition (SVD) [Funk, 2006].

In [Wang et al., 2010], a tag-based neighborhood method is incorporated into a traditional collaborative filtering approach. Tag information generated by online users is used to retrieve closer neighbors for the target user and the target item. The underlying idea of their neighborhood method is to combine both usual rating neighbors and tag neighbors to find the best neighbors. The evaluation on the tag-enhanced MovieLens data set shows that such an approach can produce more accurate rating predictions compared to other algorithms based on non-negative matrix factorization and Singular Value Decomposition. In particular, the observed improvements in predictive accuracy were comparably strong for sparse data sets as more data sources are used.

In [Sen et al., 2009b], Sen et al. propose tag-based recommender algorithms which they call "tagommenders". The idea is to utilize tag preference data in the recommendation process in order to generate better recommendation rankings than state-of-the-art baseline algorithms. Since no tag preference data is available, the tag preferences of the target user have to be estimated before the algorithm can predict a user's preference for the target item. To that purpose, the authors evaluate a variety of tag preference inference algorithms. Such algorithms estimate the user's attitude toward a tag, that is, if and to which extent a user likes items that are annotated with a particular tag. Their results show that a linear combination of all preference inference algorithms performed best, that is, algorithms that exploit a variety of signals such as implicit and explicit user data work best.

After that, the rating prediction for an item is based on the aggregation of the inferred user preferences for the tags assigned to that item. Again a hybrid approach achieved the best accuracy results, followed by an SVM-based method. Overall, the evaluation on a tag-enhanced MovieLens data set shows that tag-based recommender algorithms utilizing users' estimated tag preferences lead to more precise recommendations than the best tag-agnostic collaborative filtering algorithm.

In [Harvey et al., 2010], a tag recommender is presented which is based on a new probabilistic latent topic model influenced by LDA [Blei et al., 2003]. The authors extend the LDA model such that information about the users who provided each annotation is taken into account leading to a tripartite topic model (TTM) which covers the whole tripartite structure of a folksonomy. The results show that their TTM approach outperforms the basic LDA approach and other popularity-based approaches on different accuracy metrics. In particular, the results are strong for sparsely annotated items which are often the case in real-world tagging systems.

In line with previous work, we present in the Chapters 3 and 4 new tag-based algorithms which are able to outperform other state-of-the-art recommendation algorithms on various dimensions such as predictive accuracy or prediction time.

2.3.5   Tag-based explanations

Tagging data is not only a means to enhance existing recommender algorithms but it can also serve as a means to strengthen and improve explanations for recommendations provided by a recommendation engine. Explanations for recommendations are one of the current research topics in the recommender system research area. They play an increasingly important role as they can significantly influence the way a user perceives the system [Tintarev and Masthoff, 2007a]. Explanation interfaces can be seen as an important part of a recommender system's user interface which is of high importance for the user's perceived quality of a recommender system [Cremonesi et al., 2011].

In the social media recommendation framework of [Guy et al., 2010b] described above, each recommended social media item is also accompanied by an explanation which includes both the users and the tags upon which the recommendation was based on. The authors provide a two-level explanation for each recommended item. On the first level, the related tags and/or users are visualized depending on the applied recommendation approach. On the second level, when the user inspects the related tag or user and moves the mouse over it, its relationship to the recommended item and to the target user is shown. The results indicate that when explanations are included in their hybrid people- and tag-based PTBR recommenders, they can lead to slightly better results than compared with the tag-based only recommender TBR. However, if explanations are excluded, TBR outperforms PTBR which shows the effectiveness of (people-based) explanations.

In [Vig et al., 2009], tag-based explanation interfaces which the authors call "tagsplanations" are described and evaluated. Vig et al. propose explanation interfaces which use *tag relevance* and *tag preference* as two key components. Tag relevance measures the strength of the relationship of the tag to the item, while tag preference indicates the strength of the relationship between a user and the tag. Consider, for example, the tag "love" for a given user-item (movie) pair. Tag preference measures how well the tag "love" describes the particular movie, while tag preference indicates the user's interest in movies about love, that is, how much the user likes/dislikes movies about love in general, independent from a particular movie.

Figure 2.9: RelSort interface for the movie *Rushmore* [Vig et al., 2009]. Note that the number of tags is truncated to 7 in this figure.

Vig et al. report the results of a within-subjects study in which they evaluated four tag-based explanation interfaces. Each interface shows a list of up to 15 tags with their corresponding tag relevance and/or tag preference values. Interface 1 shows both relevance and preference values and sorts the tags by relevance (RelSort, see Figure 2.9), while Interface 2 sorts the tags by preference (PrefSort). Interface 3 only shows relevance and sorts tags by relevance (RelOnly), while the last Interface 4 only shows preference and sorts the tags accordingly (PrefOnly).

The results of their questionnaire show the importance of the concepts tag relevance and tag preference in tag-based explanations. They both help to increase justification (helping users understand their recommendation), effectiveness (helping users decide whether they will like the item), and mood compatibility (helping users determine whether an item fits their current mood). In particular, RelSort (Figure 2.9) performed best in all three evaluation dimensions.

Note that only a few works exist that see tagging data as a means to explain recommendations. In the Chapters 5 and 6, we present and evaluate new explanations based on tags.

2.3.6 Outlook

In recent years, exploiting tagging data for recommendations has become an active research topic in the field of recommender systems. Tag-based computing can further improve the quality of recommender systems and leads to new possibilities but also to a number of new research questions.

One of the main research questions which are currently being addressed in the recommender system community is the problem of how to make recommendations for a group. Group recommendations play an increasingly important role in some domains such as the movie and restaurant domains where users share and experience the recommended items often together with other users. However, only a few works deal with group recommendations which take the knowledge of other group members' opinions and the interpersonal social influence among the group members into account [O'Connor et al., 2001; Yu et al., 2006; Shang et al., 2011]. Note that the interpersonal influence is not to be confused with user similarity, which is content-dependent, while interpersonal influence is content-free [Huang et al., 2010]. The interpersonal influence is triggered by social relation rather than item content. Think, for example, of a fan who accepts any recommendations from his favorite football player, even though the recommended items might be of low value to him.

In [Shang et al., 2011], for example, a social contagion model that takes the social influence of the group members into account, is presented. A spreading algorithm is used to iteratively calculate the group's opinion about an item. The social network is modeled as an undirected graph in which nodes represent users and edges represent friendship relations between users. Recently, however, asymmetric social networking platforms such as Google+ and Twitter[18] have emerged which allow us to model the social network as a directed graph. Such a model can provide a more fine-grained view of the available relations in a social network.

According to [Jameson and Smyth, 2007] basically three group recommendation approaches are widely used in the literature: (1) merging the recommendation sets of each group member (see, for example, the POLYLENS system [O'Connor et al., 2001]), (2) combining each group member's rating prediction into one prediction (think, for example, of a simple aggregation method which bases its prediction on the lowest rating prediction in a group [O'Connor et al., 2001]), and (3) building a preference model for the group by taking the preferences of each group member into account (see, for example, the group preference model of [Yu et al., 2006]).

There exists a close relationship between group recommendations and social recommendations where the interpersonal social influence between users is also very important. In social recommendations, a person recommends an item to another person which can be modeled by a ternary relationship (*sender, receiver, item*). Note that traditional recommender systems only take the pair of (*receiver, item*) into account ignoring the influence of a third person. However, social recommendations play an important role in our everyday lives. For example, if a boy recommends a dress to his girlfriend, she will probably accept the recommendation because the recommendation comes from the boyfriend. On the other hand, if the recommendation would come from another person, e.g., the mother of the girl, the impact of the recommendation can decrease significantly [Huang et al., 2010]. A social utility function is introduced in [Huang et al., 2010] which measures the usefulness of a social recommendation by aggregating all three aspects – sender's influence, receiver's interest, and item quality – in the joint value of the Hadamard product. However, the authors do not make use of tagging data which we see as a challenging and interesting research direction for social recommendations.

Furthermore, we believe that future work will concentrate on topics of bringing semantics to tagging data (see, for example, [Xu et al., 2011a] and [Cattuto et al., 2008]). Semantically enhanced tags will further improve various aspects of recommenders such as accuracy, diversity, or explanation facility.

In general, we see tagging data as a bridge between different technologies and concepts such as the Semantic Web and the Social Web [Passant, 2007] or search and recommendation [Noll and Meinel, 2007]. In tagging data we see a helpful means to vanish the existing borders between well-established technologies and concepts. Tagging data will close the existing gaps in future work.

---

[18]http://plus.google.com, http://twitter.com

Chapter 3

LocalRank – A graph-based tag recommender

Tag recommenders are designed to help the online user in the tagging process and suggest appropriate tags for resources with the purpose to increase the tagging quality [Jäschke et al., 2008]. In recent years, different algorithms have been proposed to generate tag recommendations given the ternary relationships between users, resources, and tags, see, for example, [Rendle et al., 2009; Rendle and Schmidt-Thie, 2010] or [Gemmell et al., 2010]. Many of these algorithms, however, suffer from scalability and performance problems, including the popular *FolkRank* algorithm [Hotho et al., 2006]. For example, even when using only a small excerpt of a commonly used social bookmarking data set, FolkRank requires about 20 seconds on a typical desktop PC (AMD Athlon II Dual Core, 2.9Ghz, 8GB Ram) to compute a single recommendation list. The question of scalability and the time needed for computing the recommendations is therefore a major issue for the different tag recommendation approaches. In this chapter, we propose a neighborhood-based tag recommendation algorithm called *LocalRank*, which in contrast to previous graph-based algorithms only considers a small part of the user-resource-tag graph. An analysis of the algorithm on a popular social bookmarking data set reveals that the recommendation accuracy is on a par with or slightly better than FolkRank while at the same time recommendations can be generated instantaneously using a compact in-memory representation.

3.1 Introduction

In recent years, community-created *folksonomies* have emerged as a valuable tool for content organization or retrieval in the Social Web [Peters and Becker, 2009]. However, the value of the community-provided tags can be limited because no consistent vocabulary may exist as users have their own style and preferences which tags they use and which aspects of the resource they annotate. A picture of a car could for instance be annotated with tags such diverse as "red", "cool", or "mine" [Jannach et al., 2010]. In [Sen et al., 2007], for example, Sen et al. report that only 21% of the tags in the MovieLens system[1] had adequate quality to be displayed to the user.

One way to counteract this effect is to provide the user with a list of tag recommendations to choose from [Jäschke et al., 2008]. When the users are provided with a set of tag suggestions, the goal is that the annotation vocabulary as a whole becomes more homogenous across users and that in addition the tagging volume increases, see [Begelman et al., 2006]. In recent years, several approaches to building such *tag recommenders* have been proposed [Krestel et al., 2009; Bundschus et al., 2009; Hu et al., 2010; Harvey et al., 2010; Rendle and Schmidt-Thie, 2010]. In this chapter, we present a novel graph-and-neighborhood-based tag recommender called *LocalRank* which is based on the ideas of the popular FolkRank algorithm. We show that LocalRank can generate tag recommendations very quickly also for larger data sets and that its accuracy is comparable to that of FolkRank on the commonly-used Delicious

---

[1]http://www.movielens.org

data set[2].

This chapter is organized as follows: Section 3.2 outlines a selection of recent developments in the field of tag recommender systems. The section also provides a detailed description of the FolkRank algorithm. In Section 3.3 we present our tag recommendation algorithm LocalRank. In Section 3.4 the experimental setup is described in more detail and a discussion of the obtained results is provided. Section 3.5 finally summarizes the main findings and gives an outlook on future work.

3.2 Tag recommendations

Following the definition of [Mika, 2007] and based on the terminology and notation of [Jäschke et al., 2007] we will first provide a formal definition of a folksonomy. Afterwards, a selection of recent tag recommender algorithms is presented.

Folksonomy

According to [Hotho et al., 2006], a folksonomy is defined as a tuple $\mathbb{F} := (U, T, R, Y, \prec)$ where

- U, T, and R are finite sets, whose elements are called users, tags, and resources,

- Y is a ternary relation between them, i.e., $Y \subseteq U \times T \times R$, called tag assignments, and

- $\prec$ is a user-specific subtag/supertag-relation, i.e., $\prec \subseteq U \times T \times T$, called subtag/supertag relation.

Note that in this work $\prec$ is an empty set. For this reason we will simply denote a folksonomy as a quadruple $\mathbb{F} := (U, T, R, Y)$.

Different tag recommendation approaches can be found in the literature which take a folksonomy as input and return personalized recommendations as output. Next we will provide a small selection of recent work in the field of tag recommendations.

Selection of recent tag recommender algorithms

Maybe the easiest way to recommend tags to users is to use *popularity-based* approaches which basically rely on the popularity of a tag [Jäschke et al., 2008; Gemmell et al., 2009b]. In [Jäschke et al., 2008], for example, the authors present different variants of the "most popular tags" method. The *most popular tags by user* method, for instance, simply returns the top n used tags for a given user, while the *most popular tags by resource* method returns the top n selected tags for a given resource. However, the accuracy of such popularity-based methods is usually very limited [Gemmell et al., 2009b] which led to the development of other more sophisticated tag recommendation approaches.

Well-known *collaborative filtering* techniques can be applied easily to recommend tags to users. In [Jäschke et al., 2007] and [Gemmell et al., 2009b], for example, the k-nearest neighbor method (kNN), as described in Section 2.1.2, is applied to compute tag recommendations.

In [Jäschke et al., 2007], the authors reduce the ternary relationship Y of a folksonomy to two 2-dimensional projections $\pi_{UR}Y$ and $\pi_{UT}Y$, where $\pi_{UR}Y \in \{0,1\}^{|U| \times |R|}$ is defined as

$$\pi_{UR}Y(u,r) = \begin{cases} 1, & \exists\, t \in T : (u,t,r) \in Y \\ 0, & \text{otherwise} \end{cases} \tag{3.1}$$

and the projection $\pi_{UT}Y$ is defined analogously as the reduction of the ternary Y relation into the two-dimensional space $U \times T$. Each projection defines a binary matrix which is then used to compute user neighborhoods. Afterwards, kNN is applied to recommend tags or resources depending on whether $\pi_{UR}Y$ and $\pi_{UT}Y$ is used.

Similarly, in [Gemmell et al., 2009b], the user is modeled as a vector over the tag space. The entries in the $U \times T$ matrix are defined by the number of times a user-tag combination occurs in a user profile. Cosine similarity is used to compute the similarity between user vectors and to calculate the neighborhood

---

[2]http://delicious.org

N of the k most similar users, under the premise that all neighbors have annotated the target resource r. The tag ranking score for a tag t given a user-resource pair (u, r) is defined as

$$w(u, r, t) = \frac{\sum_n^N sim(u, n) * d(n, r, t)}{k} \tag{3.2}$$

where distance $d(n, r, t)$ returns 1 if neighbor n has annotated resource r with tag t, and 0 otherwise. The authors also present an alternative model which focuses on the similarity between recourses instead of users and operates on the $R \times T$ matrix.

The logical next step in the literature of tag-based recommender systems was to directly exploit the ternary relationship in tagging data, see, for example, [Symeonidis et al., 2008; Rendle et al., 2009] or [Rendle and Schmidt-Thie, 2010]. As described in Section 2.3.4, the idea is to view the ternary relationship as a three dimensional tensor (cube) and to apply low rank approximations for tensors on it.

In [Hotho et al., 2006], the authors present a *graph-based* tag recommender algorithm called *FolkRank*. FolkRank is a popular graph-based recommendation approach which was inspired by Google's PageRank [Brin and Page, 1998] and which is still used as a baseline for comparison in the development of new tag recommender approaches today.

Later on, different other tag recommendation algorithms have been proposed that rely on techniques such as tensor factorization and latent semantic analysis [Rendle et al., 2009; Symeonidis et al., 2010], follow a probabilistic approach [Krestel et al., 2009; Bundschus et al., 2009; Hu et al., 2010; Harvey et al., 2010] or a clustering approach [Song et al., 2008], or use hybridization strategies [Gemmell et al., 2010]. Some approaches also even go beyond recommendation, and try to automatically generate and attach personalized tags for Web pages [Chirita et al., 2007]. Since our LocalRank algorithm presented in Section 3.3 is based on the ideas of FolkRank, we will discuss FolkRank in more detail.

FolkRank

As the name suggests, the FolkRank algorithm is based on Google's PageRank algorithm. The main idea of PageRank is that pages are important when linked by other important pages. Therefore, PageRank views the Web as a graph and uses a weight spreading algorithm to calculate the importance of the pages. FolkRank adopts this idea and assumes that a resource is important if it is tagged with important tags from important users. As a first step, a given folksonomy $\mathbb{F} = (U, T, R, Y)$ is converted into an undirected tripartite graph $\mathbb{G}_\mathbb{F}$, where the set of nodes $V = U \,\dot\cup\, T \,\dot\cup\, R$ and the set of edges E and their weights is determined by the elements of Y.

Note that the folksonomy graph $\mathbb{G}_\mathbb{F}$ is different from the directed unipartite Web graph. Hotho et al. therefore propose the Folksonomy-Adapted PageRank (FA-PR) algorithm to compute a ranking of the elements and which also takes the weights of the edges into account[3]. Since $\mathbb{G}_\mathbb{F}$ is undirected, a part of the weight spread over an edge will flow back in each iteration.

Formally, the weight spreading function is $\vec{w} = dA\vec{w} + (1 - d)\vec{p}$, where A is the row-stochastic[4] version of the adjacency matrix of $\mathbb{G}_\mathbb{F}$, $\vec{w}$ is the vector containing the rank values for the elements of V, $\vec{p}$ a preference vector whose elements sum up to 1 and d a factor determining the influence of $\vec{p}$. When a non-personalized ranking of the elements of $\mathbb{G}_\mathbb{F}$ is computed, d can be set to 1. When the goal is to personalize the ranking (or support topic-specific rankings), more weight can be given to elements in $\vec{p}$ which correspond to the user preferences or a given topic. Similar to PageRank, Folksonomy-Adapted PageRank works by iteratively computing $\vec{w}$ until convergence is achieved.

The FolkRank algorithm finally computes $\vec{w}$ two times – one time including the user preferences and one time without them – and compares the differences between the rankings of the two $\vec{w}$ vectors. The "winners" of the inclusion of the preference vector therefore get higher rank values. Recommending tags for a given resource or user can be accomplished by taking the n elements with the highest rank values.

Overall, FolkRank has shown to lead to highly accurate results and even the more recent algorithms mentioned above are only slightly more accurate than FolkRank on some evaluation data sets. However, one of the major issues of FolkRank are the steep computational costs involved in the computation of

---

[3]Note that FolkRank is not limited to the calculation of weights for the tags but can also be used to compute weights of users and resources.

[4]The rows of the matrix are normalized to 1 in the 1-norm.

recommendations. Note that while the non-personalized ranks can be computed in an offline phase, this is not possible for the personalized ranking. To get an estimate of the computational costs we conducted an experiment based on the original Java implementation provided by the developers of FolkRank[5] and evaluated it on three Delicious data sets at different density levels. Computing a single recommendation list for this data set consisting of about 36,000 thousand users, 70,000 bookmarks, 21,000 tags, and 7,000,000 assignments required about 20 seconds on a typical desktop PC (AMD Athlon II Dual Core, 2.9Ghz, 8GB Ram) when the maximal number of iterations is set to 10. Note that the maximal number of iterations is relatively low compared to the numbers reported in [Hotho et al., 2006], e.g., 39. When pre-computing the unbiased ranks, the running time is reduced to about 10 seconds on average. Since FolkRank always propagates the weights through the whole network, the non-personalized weights have to be re-computed (or at least updated on a regular basis) when new tag assignments are added to the system.

The question of scalability and the time needed for computing the recommendations is a major issue for the different tag recommendation approaches. In [Rendle et al., 2009] and [Symeonidis et al., 2010], for example, the authors conclude that FolkRank does not scale to larger problem sizes and report much shorter running time figures for their own tensor factorization approach. Another approach to handle the scalability problem was developed in [Song et al., 2008], who use a clustering approach to allow for "real-time" recommendation.

In this chapter we also focus on the issue of scalability of tag recommendation to larger data sets. We propose a graph-and-neighborhood-based tag recommendation approach, which is not only capable of generating tag recommendations very quickly also for larger data sets, but which can also be efficiently updated when new data arrives. At the same time, we show that despite its simplicity, the accuracy of our method is comparable to that of FolkRank for the commonly-used Delicious data set.

In the next section, we will present our novel algorithm called LocalRank which is based on the ideas of FolkRank.

3.3   LocalRank

In contrast to FolkRank, LocalRank computes the rank weights based only on the local "neighborhood" of a given user and resource. Instead of considering all elements in the folksonomy, LocalRank focuses on the *relevant* ones only. Given a folksonomy $\mathbb{F} = (U, T, R, Y)$, its representation as $\mathbb{G}_{\mathbb{F}}$, a user $u \in U$, and a resource $r \in R$, we first compute the following sets of relevant elements as follows:

- $Y_u \subseteq Y$ is the set of all (u, t, r)-assignments of Y where u is the given user.

- Analogously, $Y_r \subseteq Y$ is the set of all (u, t, r)-assignments of Y where r is the given resource.

- The set of user-relevant tags T_u is defined to be the set of all tags appearing in the (u, t, r)-assignments of Y_u.

- The resource-relevant tags T_r are defined analogously as the set of tags from the assignments in Y_r.

- The overall set of relevant tags to be ranked by the algorithm is $T_u \cup T_r$.

Figure 3.1 visualizes the local neighborhood of a user and a resource as two subgraphs of $\mathbb{G}_{\mathbb{F}}$, constructed using the sets Y_u and Y_r. The side aspect is that the sets can be represented efficiently as a compact data structure in memory. Note that the two subgraphs can also be connected in $T_u \cap T_r$, i.e., the same tag can occur in both subgraphs.

The rank computation in LocalRank takes into account how often a certain tag was used by a user and how often a tag was attached to a resource. A similar approach was presented as *most popular tags by user* and *most popular tags by resource* in [Jäschke et al., 2008]. Although the efficiency of the combination of these approaches – known as *most popular ρ-mix* – is comparable to our approach, the accuracy results, however, are worse than those of FolkRank. Note that in our approach the popularity information is used as a factor in the rank computation of each tag in $T_u \cup T_r$.

---

[5]http://www.kde.cs.uni-kassel.de/code

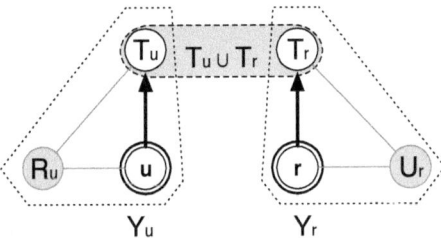

Figure 3.1: Neighborhood of relevant tags for a given user-resource query.

Rank computation and weight propagation in LocalRank is done similar to FolkRank but without iterations. The arrows in Figure 3.1 indicate the direction of the propagation of user and resource weights (see below) towards the tags.

In the FolkRank implementation the weight of a node v depends on the total number of nodes $|V|$ in the folksonomy and is set to $w = 1/|V|$. The frequency of the node's occurrence in Y is denoted as $|Y_v|$ and is defined as the number of (u, t, r)-assignments in Y in which v appears. Overall, in FolkRank, the amount of weight spread by a node v to all its adjacent nodes is $w/|Y_v|$.

LocalRank, in contrast, approximates the weights for a given user u and resource r with $w = 1/N$, where N is the total number of their neighbors in $\mathbb{G}_\mathbb{F}$. The amount of weight that is spread by the user and resource is calculated as $w/|Y_u|$ and $w/|Y_r|$ respectively.

In $\mathbb{G}_\mathbb{F}$, both algorithms calculate the weight gained by a node x by multiplying the spread weight $w/|Y_v|$ with the weight of the edge (v, x) which is equal to $|Y_{v,x}|$. While FolkRank repeatedly computes the weight gained by x for each (v, x) pair of nodes, LocalRank computes it once for each tag t in $T_u \cup T_r$. The rank of each tag $t \in T_u$ is calculated as follows:

$$rank(t) = |Y_{u,t}| \times \frac{1/N}{|Y_u|} \qquad (3.3)$$

The rank of tags in T_r is calculated similarly:

$$rank(t) = |Y_{r,t}| \times \frac{1/N}{|Y_r|} \qquad (3.4)$$

Intuitively, we finally assume that tags that appear in both sets ($t \in T_u \cap T_r$) are on principle more important than the others and should receive a higher weight. Therefore we sum up the individual rank weights obtained from the two calculations:

$$rank(t) = |Y_{u,t}| \times \frac{1/N}{|Y_u|} + |Y_{r,t}| \times \frac{1/N}{|Y_r|} \qquad (3.5)$$

LocalRank propagates the weight of the given user and resource nodes to all their adjacent tags. Therefore, it computes rankings for user and resource relevant tags and returns a list of tags and their ranks. Tag recommendations are generated by picking the top n elements with the highest rank values.

Note that in our evaluation we also experimented with a variation of the calculation scheme in which we introduced a weight factor to balance the importance of the different tag sets. The intuition behind this idea was that tags in T_r are generally more important than those in T_u because they describe the resource. Elements of T_u capture the popularity of a tag with the particular user and should have less importance as they are not necessarily meaningful to the resource. A similar approach to balancing the influence of user and resource related tags was presented in [Jäschke et al., 2008]. We tested different weight factors systematically[6], but the experiments, however, showed that the introduction of such a weight factor did not help to further improve the results.

---

[6]The weight factor was varied from 0 to 1 in steps of 0.1.

3.4 Evaluation

3.4.1 Data sets

In order to evaluate our approach both with respect to accuracy and run-time behavior, we ran tests on different versions of the Delicious data set, which is also used by many other researchers in the area of data mining and tag recommendation.

Delicious is a "social bookmarking tool", where users can manage collections of their personal Web bookmarks, describe them using keywords (tags) and share them with other users. For our experiments, we used a data set of users, bookmarks, and tags provided on courtesy of the DAI-Labor[7], which in its raw version contains more then 400 million tags applied to over 130 million bookmarks by nearly 1 million users.

In order to compare this work with previous work, we first extracted a smaller subset of manageable size from the large data set which included only the tag assignments posted between July 27 and July 30, 2005. By recursively adding tag assignments posted prior to July 27 for all users and resources present in the subset, a "core folksonomy" was constructed (as was also done in [Jäschke et al., 2007]). After this initial extraction step, we also applied p-core preprocessing to the data set. This preprocessing step guarantees that each user, resource, and tag occurs in at least k posts. That way, infrequent elements are removed from the folksonomy, thus reducing potential sources of noise in the data. At the same time, the *density* of the data is increased. Varying the p-core level therefore helps us to analyze the predictive accuracy of our methods at different density levels. In summary, experiments have been run on the three p-core levels 1, 5, and 10 (see Table 3.1 for an overview). As suggested in the literature we removed for the p-core 5 and p-core 10 data sets all posts that had more than 30 tags, as they usually are spam.

	p-core 1	p-core 5	p-core 10
Users	71,756	48,471	36,486
Tags	454,587	47,984	21,930
Resources	3,322,519	169,960	70,412
Y-assign.	17,802,069	8,963,895	7,157,654

Table 3.1: Data sets used in experiments.

3.4.2 Evaluation procedure

We used the *LeavePostOut* evaluation procedure described in [Jäschke et al., 2007], a variant of leave-one-out hold-out estimation. For all preprocessed folksonomies, we first created a subset $\tilde{U}$ consisting of 10% randomly chosen users from U (the test set). For each user in $\tilde{U}$, we picked one of the user's posts randomly. A post p is a tuple $(u, r, tags(u, r))$, where $tags(u, r) := \{t \in T \mid (u, t, r) \in Y\}$ is the set of tags associated with the post. The task of the tag recommender consists of predicting a set of tags $\tilde{T}(u, r)$ for p based on the folksonomy $\mathbb{F} \setminus \{p\}$. The predictive accuracy was determined using the usual information retrieval metrics *precision* and *recall*:

$$precision(\tilde{T}(u, r)) = \frac{|tags(u, r) \cap \tilde{T}(u, r)|}{|\tilde{T}(u, r)|} \tag{3.6}$$

$$recall(\tilde{T}(u, r)) = \frac{|tags(u, r) \cap \tilde{T}(u, r)|}{|tags(u, r)|} \tag{3.7}$$

The F1 metric, finally, was computed as the harmonic mean of precision and recall. The size of $\tilde{T}(u, r)$, that is, the length of the recommendation list, influences precision and recall. Longer recommendation lists naturally lead to higher recall values and lower precision. In the experiments, we therefore varied the length of the recommendation lists n from 1 to 20. Note that for the p-core level 1 folksonomy and also for the p-core level 5 folksonomy, the average number of tags per resource is below 20 (3 for p-core

---

[7]http://www.dai-labor.de/en/irml/datasets/delicious

1, 17 for p-core 5), which means that a precision of 100% cannot always be achieved, for example, when n is set to 20.

We used the following other parameters in our experiments. For FolkRank, we used the parameters suggested in [Jäschke et al., 2007] and set the weight parameter d to 0.7. The parameter ϵ is used in FolkRank as an indicator of reaching convergence. This means that no further iterations were made and the results were returned when the sum of all weight changes was less than 10^{-6}. As suggested in [Jäschke et al., 2007] we set the maximum number of iterations to 10 as an alternative stop condition.

3.4.3 Accuracy results

Figures 3.2 to 3.4 show the accuracy results for the different p-core levels. On the left hand side of the figures, we plot precision and recall values for the different recommendation list lengths. At the right hand side, the values of the F1 measure are shown for recommendation lists of varying length.

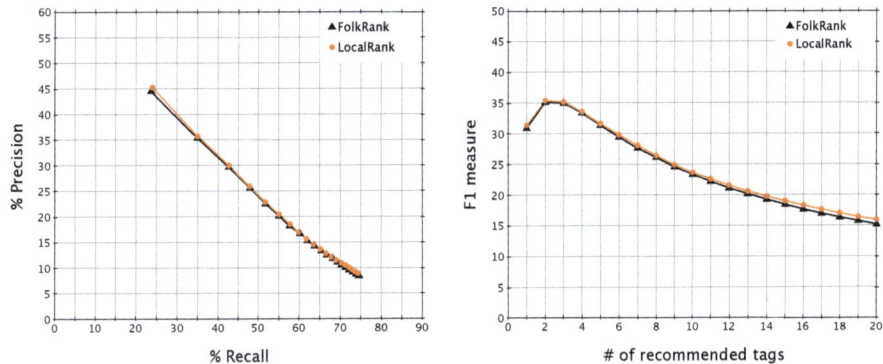

Figure 3.2: Results for the p-core level 1 data set.

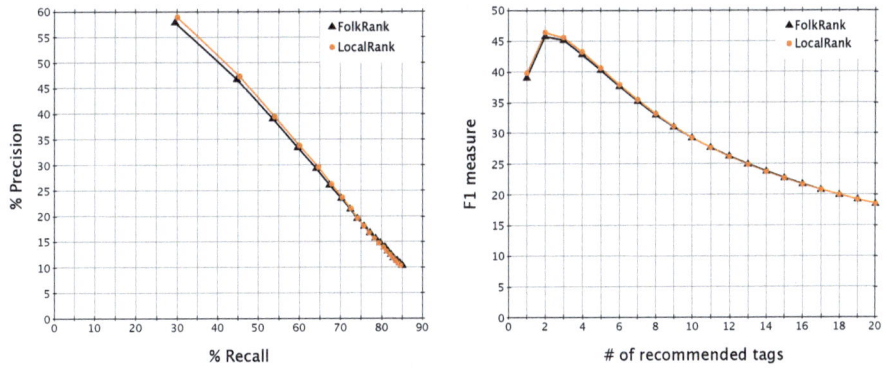

Figure 3.3: Results for the p-core level 5 data set.

Regarding the F1 measure, no strong differences between FolkRank and our LocalRank metric can be observed for all data sets. On the p-core 1 data set, LocalRank is slightly better on the overall F1 measure. A closer look reveals that LocalRank achieves higher precision and recall values for list lengths of $n > 11$. LocalRank also leads to slightly better values than FolkRank with respect to both measures

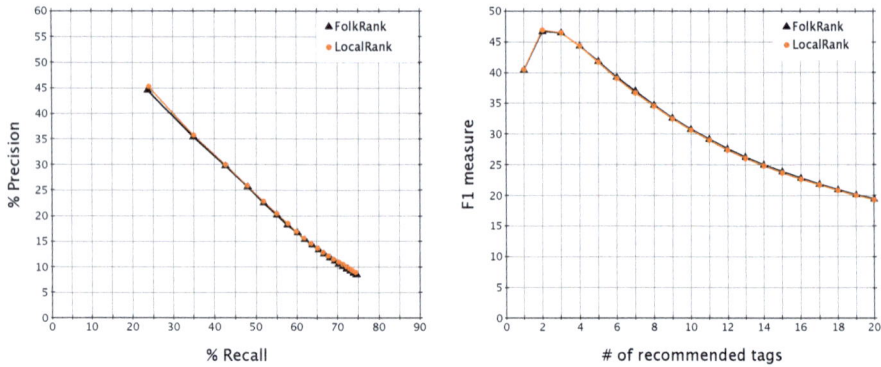

Figure 3.4: Results for the p-core level 10 data set.

for the p-core 5 data set (Figure 3.3) and for list lengths $n < 8$. The results for the p-core 10 data set are nearly identical for all evaluated recommendation list lengths, see Figure 3.4.

We conducted a sign test to analyze whether the observed differences are statistically significant using the method suggested in [Demšar, 2006]. For the p-core 5 and p-core 10 data sets, no significant differences regarding the obtained F1 measure for the two algorithms could be observed for all list lengths. For the largest and most realistic p-core 1 data set, however, LocalRank's F1 values are significantly higher ($p < .05$) for list lengths greater than 11. Overall, we therefore conclude that LocalRank is mostly on a par with FolkRank with respect to predictive accuracy on the Delicious data set at the examined p-core levels and even outperforms FolkRank in certain situations on low-density data sets.

We are aware that in very recent works new algorithms have been proposed which outperform FolkRank's predictive accuracy on certain data sets, collected for example from BibSonomy[8]. Gemmel et al. in [Gemmell et al., 2010], for instance, evaluate their hybrid approach on a p-core 20 data set collected from Delicious and observed an improvement over FolkRank. This more recent and very dense data set (p-core 20), which also involved manual selection of users and tags, was not available to us so that a direct comparison was not possible. Rendle et al. in [Rendle and Schmidt-Thie, 2010] compare their tensor factorization approach with FolkRank on a very small BibSonomy data set and could show that for longer recommendation lists their approach is slightly better on the F1 measure. Therefore, we view FolkRank still as one of the state-of-the-art techniques for tag recommendation and use it as a baseline for comparison because most current literature refers to it as a baseline. The availability of the source code is also a reason to chose FolkRank in order to ensure a fair comparison between algorithms.

3.4.4 Run-time efficiency results

As mentioned above, because of FolkRank's approach to propagate weights over the full folksonomy for each query, the algorithm suffers from scalability problems which are mentioned also in [Rendle et al., 2009] and [Gemmell et al., 2010].

Time measurements. Table 3.2 shows the average time needed for generating one recommendation list for the different p-core levels of the Delicious data sets. Note that with the Java-based version of the original FolkRank implementation from [Hotho et al., 2006], more than 20 seconds are required for generating one single recommendation list using the above-described hardware configuration. As described in Section 3.2, FolkRank computes the rank vector $\vec{w}$ using the Folksonomy-Adapted PageRank (FA-PR) two times: with and without the preference vector. The first two columns of Table 3.2 show the computation time needed for these two phases. When we assume that the folksonomy does not change,

---

[8]http://www.bibsonomy.org

	FA-PR w. preferences	FA-PR w/o preferences	FolkRank total	LocalRank
p-core 1	18,774	20,336	39,110	< 1
p-core 5	15,320	16,959	32,279	< 1
p-core 10	9,390	10,466	19,856	< 1

Table 3.2: Running times for recommendations in milliseconds.

the non-biased preference weights can be computed in advance and do not have to be re-computed for each recommendation. When relying on this re-use the computation time for FolkRank can be cut by about 50%.

Implementation and memory requirements. Similar to the implementation of FolkRank, our implementation of LocalRank is memory-based, that is, all the required data is kept in memory. Actually, the time needed for the calculation of a recommendation list is on average constantly below one millisecond and does not increase when the size of the folksonomy increases. Beside the lower computational complexity of the neighborhood-based LocalRank algorithm itself, the more or less constant access time is made possible through a compact in-memory representation of the data and a pre-processing step at startup. In the pre-processing step, simple statistics such as $|Y_u|$, $|Y_r|$, and the number of neighbors for each user and resource are pre-computed. In addition, two adjacency lists are constructed that represent the graph structure and are required for the weight propagation step: One stores the information which user posted which tags, the other one contains information about the tags attached to each resource. Once the pre-processing step is performed, the generation of recommendation lists at run-time is based on simple arithmetical operations based on the data which are organized in lookup tables. Note that when new data comes in, the lookup tables can be very quickly updated because only local changes in the "neighborhood" of the newly added elements have to be made.

The required overhead in terms of additionally required memory is limited. For the simple counting statistics (e.g., number of assignments per tag) 4 integer arrays with a total size of $2 * |U| + 2 * |R|$ are required. Two further hash maps are used to store the weights $|Y_{u,t}|$ and $|Y_{r,t}|$ of existing user/tag and resource/tag combinations in $|Y|$. Finally, the two adjacency lists are of length $|U|$ and $|R|$, where each list entry points to its assigned tags, the total number of which is $|T|$. Overall this means that $|Y|$ pointers to elements of T are required.

Comparison with other approaches. Based on our compact in-memory representation, even the p-core 1 data set can be kept in memory. Note that for example in the work by [Gemmell et al., 2010] "due to memory and time constraints" only a 10% fraction of a given Delicious data set is used. This data set is by the way the largest one in their evaluation and with 700,000 tag assignments, which is more than twenty times smaller than the p-core 1 data set used in our experiments. Note that for even larger data sets, one additional implementation option for LocalRank would be to store the most memory-intensive adjacency lists on disk in a (NoSQL) database. Typical database lookups with the given hardware configuration and data volumes usually take a few milliseconds per query. The prototypical implementation of a disk-based recommender for very large folksonomies is still an open issue in the field of tag recommender systems.

Another work which reports prediction run times is [Rendle et al., 2009]. Here, Rendle et al. compare the run times of their tensor factorization approach with FolkRank. After a linear time learning phase, their algorithm makes predictions based only on the learned model. The needed prediction time depends only on the relatively small number of factorization dimensions for users, resources, and tags as well as the number of tags $|T|$. A characteristic of their method is that it achieves better accuracy results when the model contains more dimensions (64 and 128) but is not accurate as FolkRank when the number of dimensions is lower (e.g., 8 or 16). In their paper, a graphical illustration with no exact number of running times is given. Running times range from nearly zero for the low-dimensional case up to about 10 or 15 milliseconds for the 64-factor model. Unfortunately, no numbers are given for the most accurate 128-dimensional model. While their implementation based on Object-Pascal very clearly outperforms their C++ implementation of FolkRank, the data sets taken from BibSonomy and last.fm[9] used in their evaluation are comparably small (2,500 and 75,000 assignments). The number of assignments in $|Y|$ used in their experiments is less than a 1% of our data sets. Unfortunately, also no information about the time needed to train the model (in particular for the higher-dimensional case) is given. Overall, while some accuracy improvements over our LocalRank method can be achieved using the approach described in [Rendle et al., 2009] when a high-dimensional model is learned, it remains partially unclear how their approach scales to larger problem sizes both with respect to training time and prediction time.

In [Song et al., 2008], a clustering-based, probabilistic approach for "real-time tag recommendation" is proposed and evaluated on data sets derived from Delicious and CiteULike[10]. The approach is based on a two-stage framework consisting of a learning phase and an online tag recommendation phase. The authors report running times of about a bit more than 1 second that are required to determine suitable tags for a given document on a server machine with 3GHz. Compared to our evaluation, their data set obtained from Delicious is small (215,088 tags) when compared to the 454,587 tags used in our p-core 1 data set. Unfortunately, the authors do not compare the accuracy of their approach with the one of FolkRank but with a relatively simple method based on vector similarity.

3.5   Summary and outlook

In this chapter, we proposed LocalRank, a runtime-efficient tag recommender algorithm, which despite its simplicity is capable of generating highly-accurate tag recommendations in real-time and even slightly outperforms FolkRank on the Delicious p-core level 1 data set. Compared to other approaches, LocalRank is not only quicker but also allows us to process larger data sets. Finally, from a practical perspective, our algorithm is also very easy to implement.

The original LocalRank algorithm is based only on local neighborhood information. In [Ulusoy, 2012], it is analyzed whether global tag information can be utilized for LocalRank. The author proposes GlobalRank and a hybrid solution called GlocalRank. The *GlobalRank* algorithm returns the most popular tags, while the hybridization strategy *GlocalRank* merges its results with the rank values of LocalRank. Note that in GlocalRank a parameter α is used to control the trade-off between GlobalRank and LocalRank. The results show that the idea of utilizing global tag information does not improve the accuracy results of LocalRank presented in this chapter.

However, perspectives for further improvements are quite a lot, among them is the development of a disk-based implementation of the algorithm, e.g., based on a database system, in order to analyze how massive tagging data can be processed in an efficient and scalable manner. Algorithm variants can also be developed in which the "depth" of the weight-spreading process can be increased, for example to the second or third level, without increasing the prediction times too much. Furthermore, the limitations of this work which are described in detail in Chapter 7 can be addressed in future work. For example, LocalRank can be analyzed on further data sets in order to determine whether it is sufficient also for other social tagging platforms to consider only the neighborhood of a given user-resource recommendation query.

---

[9]http://www.last.fm
[10]http://www.citeulike.org

Chapter 4

Improving recommendation accuracy based on item-specific tag preferences

Recent research has indicated that "attaching feelings to tags" is experienced by users as a valuable means to express which features of an item they particularly like or dislike [Vig et al., 2010]. When following such an approach, users would therefore not only add tags to an item as in usual Web 2.0 applications, but also attach a preference (*affect*) to the tag itself, expressing, for example, whether or not they liked a certain actor in a given movie. In this chapter, we show how this additional preference data can be exploited by a recommender system to make more accurate predictions.

In contrast to previous works, which also rely on so-called tag preferences to enhance the predictive accuracy of recommender systems [Sen et al., 2009b; Vig et al., 2010], we argue that tag preferences should be considered in the context of an item. We therefore propose new schemes to infer and exploit context-specific tag preferences in the recommendation process. An evaluation on two different data sets reveals that our approach is capable of providing more accurate recommendations than previous tag-based recommender algorithms and recent tag-agnostic matrix factorization techniques.

4.1 Introduction

Beside the usage of tags for improved item retrieval, various ways of exploiting these additionally available pieces of information have been proposed in recent years to build more effective recommender systems [Diederich and Iofciu, 2006; Hotho et al., 2006; de Gemmis et al., 2008; Tso-Sutter et al., 2008; Zhen et al., 2009; Bogers and van den Bosch, 2009; Sen et al., 2009b; Wang et al., 2010; Xu et al., 2011b]. Overall, the goal of many tag-based recommendation approaches is to exploit the existing interactions between users, items, and tags to improve the effectiveness of the recommender system, measured, for example, in terms of the predictive accuracy or the coverage of the algorithm.

In most existing approaches to tag-based or tag-enhanced collaborative item recommendation, the main assumption is that preference information provided by the user community is only available for the items. Only recently, first ideas have been put forward that consider the possibility to attach preferences also to the tags themselves and use this information to improve different quality aspects of the recommendation process.

In [Sen et al., 2009b], for example, the goal is to leverage information about the users' *estimated preference for individual tags* to generate more precise recommendations. Assuming that no explicit tag preference information is available, the first step in their "tagommenders" is therefore to estimate the user's attitude toward the different tags. In the movie domain, this would, for example, mean to estimate if and to which extent a user likes movies that are, for example, annotated with the tag "animated". After that, the rating prediction for an item is based on the aggregation of the inferred user preferences for the tags assigned to that item. An analysis of several algorithms and preference inference metrics on a tag-enhanced MovieLens data set showed that more precise recommendations can be made when the user's *estimated* tag preferences are taken into account.

Vig et al. later on report on a first study on using *explicit tag preferences*, which they call "tag expressions" [Vig et al., 2010]. In their field study, the users of the MovieLens recommender were allowed to share tags and the associated *affect*[1] – like, dislike or neutral – to the tags attached to the movies. This way, users could express which features of a movie they particularly liked or disliked. Among other aspects, their study revealed that users particularly appreciated this new feature, a fact that was measured in increased user satisfaction. Above that, allowing users to express affect associated with tags also helped to increase the volume of the contributed tags as well as their quality.

While the work of Vig et al. shows, for example, how the users' satisfaction with the system can be increased, they do not propose any algorithms for improving the recommendation accuracy based on explicit tag preferences. Note that in contrast to Vig et al.'s "tag expressions", the tagommender algorithms proposed by Sen et al. rely on "global" tag preferences, which means that a tag is either liked or disliked by a user, independent of a specific item. Thus, a particular user either likes movies annotated with the tag *animated* or not.

In our own previous work [Gedikli and Jannach, 2010c], which was developed independently of and in parallel with Vig et al.'s study, we proposed first methods aimed at exploiting *item-specific tag preferences* to compute more precise recommendations. The intuition behind this idea was that the same tag may have a positive connotation for the user in one context and a negative in another. For example, a user might like *action movies* featuring the actor *Bruce Willis*. At the same time – being used to see this actor mainly in action movies – the user might dislike the performance of Bruce Willis in a *romantic movie*. First experiments on a tag-enhanced data set and a neighborhood-based method for estimating item-specific tag preferences revealed that the predictive accuracy can be improved when compared with a similar method that only takes global tag preferences into account [Gedikli and Jannach, 2010c].

This chapter extends our previous work in different ways. First, we show how a new metric to derive user- and item-specific tag preferences can help us to produce more accurate recommendations when incorporated in a method based on Support Vector Machines (SVM), which showed superior performance in previous work, see, for example, [Sen et al., 2009b]. Beside comparing our algorithm with the best-performing method from Sen et al., we also compare with a recent tag-agnostic matrix factorization method and our own previously presented method. In addition to experiments with the tag-enhanced MovieLens data set which was also used by Sen et al., we conducted experiments in which we varied the density of the tagging data. Above that, accuracy measurements were taken based on a new data set consisting of *real* item-specific tag preferences in order to provide a more precise picture of the potential benefits of item-specific tag preferences. Real tag preference values were collected in a user study [Gedikli et al., 2011b] focusing on the explanatory power of tag preferences in the sense of [Vig et al., 2009]. The next chapter contains a detailed description of this study.

This chapter is organized as follows: In the next section, we outline the overall preference inference and recommendation process on an illustrative example. In Section 4.3, we present a scheme for automatically inferring user- and item-specific tag preferences. Afterwards, we show how the additionally available tag preference information can be exploited to make more accurate predictions. In Section 4.5, the results of the comparative evaluation of the different methods on two different data sets are discussed. The chapter ends with a discussion of related approaches and an outlook on future work.

4.2   Illustrative example and overview of the approach

Before giving details of the algorithms, we illustrate the basic rationale of our method in the following example. Let us assume user *Bob* has attached tags to different movies and given overall ratings on a scale from 1 to 5 as shown in Table 4.1. Bob particularly likes action movies featuring Bruce Willis and romantic movies featuring Sandra Bullock, but appears to dislike romantic movies starring Bruce Willis.

A method that automatically infers global preferences or ratings for tags such as the one described in [Sen et al., 2009b] would probably derive a relatively high value for the tags "Bruce Willis" and "action". At the same time, the tag "romance" would receive a luke-warm rating somewhere between 3 and 4

---

[1]We use the notion of *preference* in this work.

Movie	Tags	Rating
M1	Bruce Willis, action, ...	5
M2	Bruce Willis, romance, ...	2
M3	Bruce Willis, action, ...	5
M4	Sandra Bullock, romance, drama, ...	5
M5	Bruce Willis, romance, drama, ...	?

Table 4.1: Tags and overall ratings of Bob.

because the user attached the tag both to a highly-liked and a disliked movie. As a result, the rating prediction for movie $M5$ based on the inferred tag preferences would may be around 4, that is, the system would tend to recommend $M5$.

Now let us assume that we knew more about the individual tags and their importance to Bob as shown in Table 4.2 (assuming that we acquired this information directly from the user). In Table 4.2, we can see that – perhaps among other reasons – the user did not like $M2$ because of Bruce Willis' appearance in a romantic movie. Since movie $M5$ is quite similar to $M2$ with respect to the attached tags, it is somewhat more intuitive *not* to recommend $M5$, which is exactly the opposite decision as in the example above.

Movie	Tags	Rating
M1	Bruce Willis (5), action (5), ...	5
M2	Bruce Willis (1), romance (2), ...	2
M3	Bruce Willis (5), action (5), ...	5
M4	Sandra Bullock (4), romance (5), ...	5
M5	Bruce Willis (?), romance (?), ...	?

Table 4.2: Tags and detailed ratings of Bob.

We therefore propose a method that is capable of making recommendations based on more detailed rating data for tags. We assume that such data will be available also in future systems, given the findings of the study of Vig et al., who showed that users enjoy sharing their affect by expressing their feelings with respect to certain features of the recommendable items [Vig et al., 2010]. In addition, we develop a method to infer this detailed tag preference data automatically for cases in which such information is not available or the data is very sparse. In the example above, we would try to approximate the detailed ratings for the items $M1$ to $M5$ as good as possible given only the overall ratings for the movies.

Next, we will present different methods to derive tag preferences from the overall ratings automatically. Afterwards, schemes to derive an overall rating prediction for a not-yet-seen item based on the ratings of its tags are proposed and evaluated. In general, our methods extend the "tagommender" algorithms and metrics proposed in [Sen et al., 2009b] in a way that they can take into account item-specific tag preferences in the recommendation process.

4.3 Estimating unknown tag preferences

If no explicit tag preferences are given, Sen et al. propose different algorithms for estimating global tag preferences from the given item ratings [Sen et al., 2009b]. According to their evaluation, the algorithm `movie-ratings` is both effective and at the same time relatively easy to implement. The algorithm is based on *tag relevance weighting*, a concept which is also used in the work on tag-based explanations described in [Vig et al., 2009].

In our evaluation, we use a simple counting metric $w(i, t)$ to measure the relevance of a tag t for an item i. The metric gives more weight to tags that have been used by users more often to characterize the item and is defined as follows[2]:

$$w(i, t) = \frac{number\ of\ times\ tag\ t\ was\ applied\ to\ item\ i}{overall\ number\ of\ tags\ applied\ to\ item\ i} \tag{4.1}$$

---

[2]Further possible metrics to determine the relevance of a tag are described in [Sen et al., 2009b].

In the `movie-ratings` algorithm, the prediction $\hat{r}_{u,t}$ of the general interest of a user u in the concept represented in a tag t, that is, the tag preference, is estimated as follows:

$$\hat{r}_{u,t} = \frac{\sum_{m \in I_t} w(m,t) * r_{u,m}}{\sum_{m \in I_t} w(m,t)} \tag{4.2}$$

In this equation, I_t corresponds to the set of all items tagged with t. The explicit overall rating that u has given to movie m is denoted as $r_{u,m}$. The general idea of the method is thus to propagate the overall rating value to the tags of a movie according to their importance.

In this work, however, we are interested in predicting the rating for a tag in the context of the target user u *and* the target item i. Note that the rating prediction in Equation (4.2) does not depend on the target item i at all. Our tag prediction function, $\hat{r}_{u,i,t}$ calculates a prediction for the target tag t, given user u and item i, as follows:

$$\hat{r}_{u,i,t} = \frac{\sum_{m \in similarItems(i,I_t,k)} w(m,t) * r_{u,m}}{\sum_{m \in similarItems(i,I_t,k)} w(m,t)} \tag{4.3}$$

Instead of considering all items that received a certain tag as done in [Sen et al., 2009b], we only consider items that are similar to the item at hand, thereby avoiding the averaging effect of "global" calculations. In Equation (4.3), the calculation of neighboring items is contained in the $similarItems(i, I_t, k)$ function which returns the k most similar items to i from I_t.

The similarity of items is measured with the adjusted cosine similarity metric. Note that we also ran experiments using the Pearson correlation coefficient as a similarity metric, which, however, led to poorer results. As another algorithmic variant, we have tried to factor in the item similarity values as additional weights in Equation (4.3). Again, this did not lead to further performance improvements but rather worsened the results.

Note that when using the user's explicit overall rating $r_{u,m}$ as in Equation (4.2), no prediction can be made for the tag preference if user u did not rate any item m tagged with t, i.e., if $I_t \cap ratedItems(u) = \emptyset$. In our previous work [Gedikli and Jannach, 2010c], we therefore also incorporated the recursive prediction strategy from [Zhang and Pu, 2007] into the tag preference prediction process, which lead to a slight performance improvement. Since such an performance improvement was, however, also observed for Sen et al.'s original methods, we will not further discuss these generally-applicable technique here in greater depth, see [Gedikli and Jannach, 2010c] for details of the evaluation.

4.4   Predicting item ratings from tag preferences

In [Sen et al., 2009b], the best-performing tag-based recommendation algorithm with respect to *precision* is a hybrid which combines the SVM-based method `regress-tag` and the tag-agnostic matrix factorization approach `funk-svd` [Funk, 2006]. In this work, we therefore propose to parameterize and evaluate the *regress-tag* method using item-specific tag preferences. Note again that these tag preferences can be explicitly available or derived as described above in Equation (4.3). In addition to that, we report accuracy results when using item-specific tag preferences for Sen et al's `cosine-tag` method, in order to study how this approach, which we proposed in our previous work [Gedikli and Jannach, 2010c], performs on additional data sets.

The regress-tag algorithm. The `regress-tag` method from [Sen et al., 2009b] is based on determining linear equations – one for each movie – which capture the possibly complex relationship between the user's tag preferences for the tags of a given item and the overall item rating. The prediction function for a user u and an item i is defined as follows, where h_0 to h_n are the coefficients of the linear equations and $\hat{r}_{u,t_i}$ from Equation (4.2) corresponds to the estimated tag preferences for the tags $t_1, ..., t_n$ attached to item i:

$$regress\text{-}tag(u,i) = h_0 + h_1 * \hat{r}_{u,t_1} + ... + h_n * \hat{r}_{u,t_n} \tag{4.4}$$

In [Sen et al., 2009b], the coefficients h_0 to h_n were chosen with the help of regression support vector machines and the *libsvm* library [Chang and Lin, 2011] because this led to the most accurate results when compared with other methods for choosing the parameters such as least-squares optimization.

We used the same approach (as well as the same software library and algorithm parameters), but different values for the tag preferences. In our algorithm variant, which we shall call `regress-tag-ui`[3], we therefore use item-specific tag preferences as described in Equation (4.3):

$$regress\text{-}tag\text{-}ui(u, i) = h_0 + h_1 * \hat{r}_{u,i,t_1} + ... + h_n * \hat{r}_{u,i,t_n} \qquad (4.5)$$

The `cosine-tag` algorithm. The `cosine-tag` method is inspired by an analogy to classical item-to-item collaborative filtering methods. The prediction of a user's rating for a given item is based on the weighted combination of the user's preference for the tags of an item. Let T_i be the set of all tags applied to an item i. The rating prediction for a user u and an item i is calculated as follows:

$$cosine\text{-}tag(u, i) = \overline{r_u} + \frac{\sum_{t \in T_i} sim(i, t) * (\hat{r}_{u,i,t} - \overline{r_{u,i}})}{\sum_{t \in T_i} sim(i, t)} \qquad (4.6)$$

The individual tag preferences are weighted according to the adjusted cosine similarity between items and tags. The similarity metric given in Equation (4.7) is used to measure the degree of consistency between the item's overall rating received by all users u who rated item i (U_i), and their predicted tag preferences for that item, that is,

$$sim(i, t) = \frac{\sum_{u \in U_i} (r_{u,i} - \overline{r_u})(\hat{r}_{u,i,t} - \overline{r_{u,i}})}{\sqrt{\sum_{u \in U_m} (r_{u,i} - \overline{r_u})^2} \sqrt{\sum_{u \in U_i} (\hat{r}_{u,i,t} - \overline{r_{u,i}})^2}} \qquad (4.7)$$

To illustrate the effect of the approach, consider the following example. Table 4.3 shows an example of inferred tag preferences for the MovieLens user John[4] and the movie *Snatch (2000)* on a 5-star scale with half-star increments. John is a real user in the MovieLens movie recommendation community. From the inferred rating data, we can see that John has particularly liked the *British* elements and tone in the movie and that John probably likes the acting of *Jason Statham* more than the acting of *Brad Pitt*, at least in this movie.

When we use Equation (4.6) to combine the estimated tag preferences into one overall rating prediction for that movie, we predict a rating value of 3.355 from John for *Snatch*. Note that John's overall (explicitly given) rating was 3. Our method in that case therefore predicts a rating value that is a little higher and closer to the next higher 3.5-star rating.

Tag	Rating	Tag	Rating
British	4.938	boxing	3.464
Guy Ritchie	4.328	cynical	3.346
crime	4.241	comedy	3.346
hilarious	4.115	Brad Pitt	3.346
Jason Statham	3.835	fighting	3.346
goofy	3.636	quirky	3.346

Table 4.3: Inferred tag preferences of the MovieLens user John for the movie *Snatch (2000)*.

4.5 Evaluation

In order to measure the predictive accuracy of the presented methods we evaluated our approach on two data sets using common experimental procedures and accuracy metrics. The results of this evaluation are described in this section. The goal of our subsequent evaluation is to analyze in which cases the predictions made by the system are more accurate than previous methods when we rely on user- and item-specific tag preferences.

---

[3] "ui" stands for *user- and item-specific* tag preferences.
[4] John is not the real name of the user. We use a pseudonym to protect the user's privacy.

Constraint		Description
Min Users/Tag	(U/T)	Minimum number of users per tag.
Min Items/Tag	(I/T)	Minimum number of items per tag.
Min Tags/User	(T/U)	Minimum number of tags a user has specified.
Min Items/User	(I/U)	Minimum number of items rated by a user.
Min Tags/Item	(T/I)	Minimum number of tags applied to an item.
Min Users/Item	(U/I)	Minimum number of users which rated an item.

Table 4.4: Quality parameters.

4.5.1   Data sets, tag quality, and data preprocessing

We used two data sets in our evaluation. First, we evaluated our methods on the "MovieLens 10M Ratings, 100k Tags" (ML) data set[5], which was also used in the analysis by [Sen et al., 2009b]. The data set consists of movie ratings on a 5-star scale with half-star increments. In addition, it contains information about the tags that have been assigned by the users to the movies. A tag assignment is a triple consisting of one user, one resource (movie) and one tag. No rating information for the tags themselves is available in the original MovieLens database. To the best of our knowledge, the 10M MovieLens data set is the only publicly available data set which contains both rating and tagging data. It contains 10,000,054 ratings and 95,580 (unrated) tags applied to 10,681 movies by 71,567 users of the online movie recommender service MovieLens.

Second, we used a new data set containing *explicit* tag preferences, which we collected in the user study on the usage of tagging data for explanation purposes reported in Chapter 5. The data set contains 353 overall ratings for 100 movies provided by the 19 participants of the study. In addition to these overall ratings, the study participants provided 5,295 explicit ratings for the tags attached to the movies. On average, every user rated about 18 movies and each movie had 15 tags assigned.

Limited tag quality is one of the major issues when developing and evaluating approaches that operate on the basis of user-contributed tags [Sen et al., 2007]. Therefore, different approaches to deal with the problem of finding quality tags have been proposed in recent years, see, for example, [Gemmell et al., 2009a], [Sen et al., 2007], or [Sen et al., 2009a].

Note that our approach of rating items by rating tags calls for a new quality requirement to tags: tags must be appropriate for ratings. For example, there is no point in attaching a rating to a tag like "bad movie" because the tag already represents a like/dislike statement. It would therefore not be clear how to interpret a preference for such a tag. In our current work and evaluation, we did not take this question into account yet, that is, we did not distinguish between tags that are appropriate for being rated and those which are not. Still, we believe that this is one key question which was not considered before and which should be taken into account in future approaches to extracting rating information for tags automatically.

For the MovieLens (ML) data set, we applied and varied the constraints shown in Table 4.4 in order to remove tags, users, or items for which not sufficient data was available. This way, we varied the quality of the existing tag information. We, for example, only considered movies, for which a minimum number of tags is assigned (Min Tags/Item). This approach was also followed in previous work. In [Vig et al., 2009], for example, the authors require that "a tag has been applied by at least 5 different users and to at least 2 different items". Additionally, content analysis methods were applied to detect redundant tags such as *violent* and *violence*, in order to replace them by one representative tag. Similar to their approach, we further automatically pre-processed the data in three dimensions by removing stop-words from the tags, by applying stemming [Porter, 1997] and by filtering out noisy tags, i.e., tags with a certain amount of characters that are not letters, numbers or spaces, e.g., elements such as smileys.

We created three different versions from the tag-enhanced MovieLens data with different constraints on data density, see Table 4.5 for an overview. Note that our quality and density requirements are relatively weak when compared, for example, with the work of [Vig et al., 2009], who required that a tag has been used by at least five users to be considered in the evaluation. As a result, the MAE values we report are in general slightly higher than those reported in [Sen et al., 2009b], who also used similar

---

[5]http://www.grouplens.org/node/73

Name	Minimum constraints						Resulting data set			
	U/T	I/T	T/U	I/U	T/I	U/I	Ratings	Users	Items	Tags
ML-high	4	4	25	20	5	5	33,973	149	668	19,976
ML-medium	3	3	20	20	5	5	77,578	179	2,952	41,821
ML-low	2	2	1	1	2	1	134,829	3,979	4,713	77,127

Table 4.5: MovieLens data set variations.

types of constraints to prune the data set.

In contrast to the MovieLens data set, our second data set from the user study (US) contains explicit tag preferences. Therefore, no estimation of the tag preferences is required. In our experiments, we, however, varied the amount of explicitly available tag preference data in order to study the differences when using only explicit ratings, only automatically inferred ratings, and a setting where one half of the rating data is real and one half only estimated. See Table 4.6 for an overview.

Name	Description
US-real	Only the explicit tag preferences provided by the study participants are used.
US-mixed	50% of the real ratings are randomly removed and replaced by estimates.
US-pred	Existing tag preferences are removed and predicted using Equation (4.3).

Table 4.6: User study data sets.

Note that for all variations of the user study data set the number of items, users, and ratings, i.e., the density levels, are the same as no further quality improvement measures were applied.

4.5.2 Algorithms

We included the following recommendation algorithms in our comparative evaluation. Our proposed schemes are:

- **regress-tag-ui**: SVM regression with user- and item-specific tag preferences as described in Equation (4.5).

- **cosine-tag-ui**: Similarity-based prediction as described in Equation (4.6).

Beside these two methods, we also experimented with the probabilistic approach to content-based recommendation proposed in [Mooney and Roy, 2000]. Instead of extracting keywords from documents as in the original approach, we simply viewed the tags of the movie as the keywords and used them to derive an estimate of whether a user will like or dislike a movie. An experimental evaluation on the ML data set, however, showed that the predictive accuracy of this simplistic approach is not competitive when compared to other techniques. This observation is in some respect in line with the observation made in [Pilászy and Tikk, 2009], in which it was also noticed that rating-based collaborative filtering schemes are more accurate than pure content- or metadata-based approaches even if the number of ratings is relatively low.

As a baseline for the comparison, we used the following algorithms:

- **regress-tag**: SVM regression as proposed in [Sen et al., 2009b] (see Equation (4.4)).

- **cosine-tag**: Similarity-based prediction as proposed in [Sen et al., 2009b].

Note that in Sen et al.'s work a hybrid algorithm which combines **regress-tag** with a matrix factorization approach performed slightly better than **regress-tag** alone. We experimented with this hybridization strategy also on our data sets, but could not reproduce their findings. Given our data, we consistently observed **regress-tag** to be the best-performing method, which we therefore use as a baseline.

Beside the two tag-aware baseline algorithms, we also measured the predictive accuracy of two tag-agnostic techniques:

- `item-item`: A classical item-to-item baseline recommendation scheme that does not exploit tag information at all. Adjusted cosine is used as a similarity function. Rating predictions are calculated as follows:

$$\hat{r}_{u,i} = \frac{\sum_{m \in ratedItems(u)} sim(m,i) * r_{u,m}}{\sum_{m \in ratedItems(u)} sim(m,i)} \tag{4.8}$$

- `funk-svd`: The recent, highly-accurate matrix factorization algorithm based on Singular Value Decomposition (SVD) proposed in [Funk, 2006], which was also used as a baseline for comparison by [Sen et al., 2009b].

We used the following algorithm parameters:

- We chose adjusted cosine as similarity metric for all schemes because experiments with other similarity metrics such as Pearson's correlation coefficient led to poorer results.

- In the `regress-tag` method, similar to the experiments reported in [Sen et al., 2009b], we set the value of c related to the error penalty to 0.005. For `regress-tag-ui` we additionally determined 10 as a suitable number of neighbors to include when calculating the item-specific tag preferences.

- In the user- and item-based `cosine-tag` scheme, the parameter k which determines the size of the neighborhood containing the k most similar items from I_t can be varied, see Equation (4.3). In order to find an optimal value we varied the parameter systematically. A neighborhood-size of 3 was determined as an optimal choice.

- For the SVD-based recommender, 30 was empirically chosen as the number of latent features.

All algorithms were implemented in our Java-based framework called *RecommenderSuite*, which also includes components to run offline experiments, do cross-validation and measure various metrics such as accuracy or coverage. The `regress-tag` methods are based on the *libsvm* implementation [Chang and Lin, 2011]; the implementation of the matrix factorization approach was adapted from the Apache Mahout project[6].

4.5.3 Accuracy metrics

We analyzed the quality of the recommendations generated by the different algorithms with the standard information retrieval metrics precision and recall presented in Section 2.2.1. We followed the evaluation procedure proposed in [Nakagawa and Mobasher, 2003] and converted the rating predictions into "like" and "dislike" statements as described in [Sandvig et al., 2007], where ratings above the user's mean rating are interpreted as "like" statements.

In each of the iterations of a cross-validation procedure, the data set was split into a training set and a test set. We then determined the set of existing "like" statements (ELS) in the test set and retrieved a top-n recommendation list of length $|ELS|$ with each method based on the data in the training set. The top-n recommendation lists were created based on the prediction score of each method. On the other hand, the set of predicted like statements returned by a recommender shall be denoted as PLS, where $|PLS| \leq |ELS|$. Based on these definitions, precision and recall values were calculated as described in Section 2.2.1.

In the evaluation procedure, recommendations and the corresponding precision and recall values were calculated for all users in the data set and then averaged. These averaged precision and recall values were then combined in the usual F1-score, where $F1 = (2 * precision * recall)/(precision + recall)$.

Beside these information retrieval metrics, we also report the usual Mean Absolute Error (MAE) numbers in order to make our results comparable with the results in the literature, in particular with Sen et al.'s work. Note that we also calculated Root Mean Squared Error (RMSE) values, but do not report these numbers here because no significant differences to the MAE values have been observed.

---

[6]http://mahout.apache.org

4.5.4   Results and discussion

Experiments on real data

Table 4.7 shows the Mean Absolute Error values for the different algorithms and the data sets from the user study in increasing order[7]. It is important to recall that the density levels of the data sets are fixed, whereas the amount of explicitly available tag preference data is varied as described in Section 4.5.1.

Algorithm	US-real	US-mixed	US-pred
regress-tag-ui	0.46	0.47	0.49
regress-tag	0.48	0.49	0.49
funk-svd	0.58	0.58	0.58
cosine-tag	0.59	0.59	0.60
item-item	0.59	0.59	0.59
cosine-tag-ui	0.60	0.59	0.60

Table 4.7: MAE results for real tag preference data.

We can see that the `regress-tag-ui` scheme proposed in this chapter to exploit user- and item-specific tag preferences leads to the smallest MAE values for all data set variations. The improvement over the previous `regress-tag` method is particularly strong when only real rating data is used. For the situation in which we only rely on estimates of item-specific tag preferences (US-pred), our method is minimally better or at least on a par with the previous method. When using only half of the existing ratings, MAE values somewhere in the middle between the two extremes can be observed. As a result, the numbers indicate that the accuracy constantly increases when more real tag preferences are entered into the system.

Another observation on this relatively dense data from the user study is that the `cosine-tag-ui` method performs slightly worse than the `cosine-tag` and even the `item-item` algorithm. For the data set US-pred, `item-item` also slightly outperforms `cosine-tag`.

Table 4.8 shows the corresponding values for precision, recall, and the F1-measure. The values were obtained using a 10-fold cross-validation on the relatively small data set.

Algorithm	US-real			US-mixed			US-pred		
	F1	Prec.	Recall	F1	Prec.	Recall	F1	Prec.	Recall
regress-tag-ui	84.07	84.66	83.50	84.43	85.03	83.86	83.30	83.90	82.73
regress-tag	83.30	83.90	82.73	83.30	83.90	82.73	83.30	83.90	82.73
funk-svd	72.39	72.99	71.82	71.97	72.57	71.41	72.39	72.99	71.82
cosine-tag	68.90	69.50	68.33	68.90	69.50	68.33	68.90	69.50	68.33
cosine-tag-ui	68.66	69.26	68.10	68.23	68.83	67.67	68.05	68.65	67.49
item-item	68.05	68.65	67.49	68.05	68.65	67.49	67.30	67.99	66.82

Table 4.8: F1, precision, and recall results for real tag preference data.

Again, the numbers show that `regress-tag-ui` slightly outperforms `regress-tag` and the other algorithms and is significantly better than the other methods. The neighborhood- and similarity-based methods show the poorest results on this metric and are also outperformed by the SVD-based algorithm. Note that there are virtually no differences in the observed numbers across the different data sets, which can be accounted to the characteristics of the dense user study data set and the chosen evaluation metric.

MovieLens data

Table 4.9 shows the MAE numbers for the different MovieLens data sets. Note that unlike the three data sets considered above the different MovieLens data sets have different density levels as described in Section 4.5.1.

---

[7]Note that we only report two decimal places in the tables due to standard error of the entries and randomness effects. However, the order of the entries is based on the third decimal place.

Algorithm	ML-high	ML-medium	ML-low
regress-tag-ui	0.64	0.64	0.69
funk-svd	0.64	0.65	0.71
regress-tag	0.65	0.65	0.69
cosine-tag-ui	0.68	0.70	0.85
cosine-tag	0.68	0.70	0.85
item-item	0.68	0.70	0.85

Table 4.9: MAE results for the tag-enhanced MovieLens data set.

Algorithm	ML-high			ML-medium			ML-low		
	F1	Prec.	Recall	F1	Prec.	Recall	F1	Prec.	Recall
regress-tag-ui	70.22	70.22	70.22	70.80	70.81	70.79	94.37	94.45	94.28
regress-tag	69.98	69.98	69.98	70.52	70.54	70.51	94.36	94.44	94.27
item-item	68.89	68.89	68.89	68.14	68.15	68.12	90.50	90.59	90.41
funk-svd	67.84	67.84	67.83	67.31	67.32	67.29	93.92	94.00	93.83
cosine-tag	66.42	66.43	66.42	64.32	64.34	64.31	90.72	90.81	90.64
cosine-tag-ui	65.47	65.47	65.46	64.03	64.04	64.01	*90.98*	*91.06*	*90.89*

Table 4.10: F1, precision, and recall results for the tag-enhanced MovieLens data set.

In these experiments, in which only estimated tag preferences are available, the funk-svd method, regress-tag, and regress-tag-ui are more or less on a par with respect to the MAE measure. The more traditional neighborhood-based methods are measurably less accurate. These findings are in line with the MAE numbers reported in [Sen et al., 2009b], where the funk-svd method was reported be even slightly better on the MAE metric by a very small margin. For the data set ML-low – the largest data set with the least constraints on the data – the performance of funk-svd is slightly worse than the regression-based approaches.

Table 4.10 finally shows F1, precision, and recall values for the three MovieLens-derived data sets, in which artificial tag preferences were used. The results show that our new scheme regress-tag-ui outperforms the best-performing algorithm regress-tag from Sen et al. on the denser data sets (ML-high and ML-medium) and is at least on a par with it on the lower-density data set ML-low. Quite interestingly, the classical tag-agnostic item-to-item algorithm performs relatively well on this metric and is even better than the matrix factorization approach on the data sets ML-high and ML-medium.

Note that the findings of our previous work, in which cosine-tag-ui was presented and compared with cosine-tag [Gedikli and Jannach, 2010c], could be reproduced, that is, that cosine-tag-ui is slightly better than cosine-tag on the F1-metric. On the denser data sets, the cosine-tag-ui method, however, does not perform as well as the other algorithms.

With respect to the absolute recall and precision, we can observe that the measured precision and recall values are mostly very similar to each other for each algorithm and data set. This is caused by the nature of the particular accuracy metric that we use in this work (see Section 4.5.3 and [Nakagawa and Mobasher, 2003] respectively)[8].

The absolute values are also significantly higher for the very sparse data set ML-low, which is also caused by the characteristics of the chosen evaluation metric, which is not designed for a comparison of the absolute numbers across such different data sets. Note that in our experiments, the data density for the real data from the user study is close to 20%. For the MovieLens data sets, we varied the density levels from 35% for ML-high over 23% for ML-medium to the very low-density data set ML-low, which has a density of 0.001%.

In summary, the main objective of this chapter is to show that accuracy improvements can be achieved when using tag preference data. Overall, the results achieved show that the usage of item-specific tag preferences can help to improve the predictive accuracy of recommender systems and that the observed

---

[8]The number of items (n) retrieved by a top-n recommender during the evaluation is user-dependent and not fixed to a constant value. This can lead to the same precision and recall values because the denominator of the precision and recall values can have the same size. This procedure is good in scenarios where the average number of positive items in the test set varies strongly for each data set as in our case.

positive effect is stronger, when the tag quality and data density increases. The evaluation on the different data sets demonstrated that especially when using real tag preference data the method `regress-tag-ui` proposed in this chapter outperforms the best-performing method from previous work. The results show further that even when all tag preferences are artificially derived from the item ratings our method is still able to slightly outperform or at least is on a par with the previous best-performing method. The predictive accuracy increases when more real data is entered into the system.

4.6 Related work and discussion

As described in Section 2.3, a plethora of methods exist which exploit tagging data for recommender systems. In [de Gemmis et al., 2008], for instance, the authors exploit tagging data for an existing content-based recommender system in order to increase the overall predictive accuracy of the system. In their approach, tags are only considered as an additional source of information used for learning the profile of a particular user. By conducting a user study with 30 users the authors show that a slight improvement in the prediction accuracy of the tag-augmented recommender compared to the pure content-based one can be achieved. In contrast, this work rather represents a collaborative filtering approach with multi-criteria ratings and is thus better capable to exploit the wisdom of the crowd to improve recommendation accuracy.

Similar to Sen et al.'s and our work, [Zhou et al., 2009] proposed a method based on probabilistic factor analysis framework that exploits the information contained in the user-item rating matrix, the user-tag tagging matrix and the item-tag tagging matrix to produce more accurate recommendations. A comparison with two recent tag-agnostic matrix factorization approaches (see [Funk, 2006] and [Salakhutdinov and Mnih, 2008]) showed that their approach leads to better RMSE values in particular when only a small part of the available data is used for training.

Note that the idea of allowing the community to rate items is an important topic not only in the area of recommender research but also in the Semantic Web community. Revyu[9] [Heath and Motta, 2007], the winner of the Semantic Web Challenge of the year 2007, is a reviewing and rating Web site which aims to aggregate review data of items (resources) on the Web. Revyu allows people to rate items by writing reviews and gives users the opportunity to add meta-data to items in the form of Web 2.0 style tags. Based on this relatively unstructured information, stronger semantics are later on derived. As stated by the authors, this functionality in itself is partially not particularly novel. The real benefit lies in the use of Semantic Web technologies and standards like RDF, SPARQL, and the principles of Linked Data [Berners-Lee, 2006] in order to expose reviews in a reusable and machine-readable format.

Note that in the Revyu system tags are merely used for classifying the reviewed items and for automatically extracting additional information. We believe that our work could complement this approach by exploiting the rating information which is implicitly contained in the tags. That way, by deriving individual preferences for the tags provided by a user, a better "understanding" of the free-text reviews could be achieved.

In contrast to works in which tags are only used to build better neighborhoods for classical collaborative filtering systems, we follow a different approach of exploiting tags for recommender systems in this chapter. In principle, we propose an approach, in which *users rate items by rating tags*, which to some extent also has a correspondence to a multi-criteria or multi-dimensional recommendation approach as described in [Adomavicius and Tuzhilin, 2001; Adomavicius and Kwon, 2007] or [Jannach et al., 2012]. In [Adomavicius and Kwon, 2007], Adomavicius and Kwon conjecture that multi-criteria ratings will play an important role for the next generation of recommender systems, in particular because multi-criteria ratings can help to handle situations in which users gave the same overall rating but had different reasons for that (which can be observed in the detailed ratings). Besides this, multi-criteria rating information can serve as a valuable source for explaining recommendations. Based on these observations, new user similarity metrics and algorithms were designed in [Adomavicius and Kwon, 2007] that exploit multi-criteria rating information leading to recommender systems of higher quality. The authors show on a small data set how exploiting multi-criteria ratings can be successfully leveraged to improve recommendation accuracy. Our approach of "rating items by rating tags" shares the advantages of these multi-criteria

---

[9]`http://revyu.com`

recommender systems such as improved accuracy and explanations. The rating dimensions are, however, not static in our approach and require metrics that are different to those put forward, for example, in [Adomavicius and Kwon, 2007].

In this respect, this work is also in line with the ideas of Shirky [Shirky, 2005], who was among the first who argued that using predefined (rating) categories leads to different challenges such as the following. First, professional experts are needed who design the rating dimensions; in addition, new rating dimensions may emerge over time that were not covered by the predefined and pre-thought static rating dimensions designed or foreseen by a domain expert. In collaborative tagging systems, the set of rating dimensions is not limited which allows users to pick their particular way of stating their preferences. Of course, this comes at the price of a less homogeneous and more unstructured set of item annotations.

To the best of our knowledge, the concept of tag preference was first introduced by [Ji et al., 2007]. The authors present a tag preference based recommendation algorithm for a collaborative tagging system. The authors first compute the target user's *candidate tag set* which consists of all tags for which a high tag preference value was predicted. Afterwards a naïve Bayes classifier is used for making recommendations by exploiting the user's candidate tag set. The proposed algorithm was evaluated on a data set collected from the social bookmarking site Delicious[10]. In contrast to the work of [Sen et al., 2009b], the tag preference predictor in [Ji et al., 2007] does not make use of item ratings at all because the Delicious data set does not support ratings for items (bookmarks) like the tag-enhanced MovieLens data set.

In [Vig et al., 2009], Vig et al. propose another concept called *tag relevance* which describes "the degree to which a tag describes an item". In the example from Section 4.2, tag relevance would measure how well the tag "Bruce Willis" describes a particular movie. Overall, in previous work *tag preference* was considered a user-specific concept whereas *tag relevance* is considered to be an item-specific concept. In contrast, in this work our proposed concept of *tag preference* is user- and item-specific which has shown to be a helpful means to capture the user's preferences more precisely and thus produce more accurate recommendations.

Beside the relation of the work in [Sen et al., 2009b], which we extend by item-specific tag preferences in this chapter, our approach is also closely related to the recent work of [Vig et al., 2010], who experimented with a recommender system interface that allowed users to assign *affect* to the tags of a movie. In [Vig et al., 2010], the authors introduce the idea of "tag expressions", which, at its heart, represents the same idea of rating items by rating tags proposed in our own previous work [Gedikli and Jannach, 2010c]. Users are able to assign a so-called *affect* (preference) – like, dislike or neutral – to tags in order to measure a user's pleasure or displeasure with the item with respect to the tag. In contrast to this work the authors are focussing on user interface aspects and how the possibility to express tag preferences affects the tagging behavior of the community. In particular, Vig et al. also analyze the design space of tag expressions and focus on three elements: preference dimensions, affect expression, and display of community affect. In this work, however, we present first algorithms that consider tag preferences (tag affects) to generate more accurate predictions which is one of the main challenges listed in the future work section of [Vig et al., 2010]. Additionally we also show how to infer the user opinion regarding a certain feature (tag) for a given item automatically.

4.7   Summary and outlook

The main new idea presented in this chapter is to incorporate item-specific ratings for tags in the recommendation process. Based on such an approach, users are able to evaluate an existing item in various dimensions and are thus not limited to the one single overall vote anymore. In contrast to previous attempts toward exploiting multi-dimensional ratings, this work aims to follow a Web 2.0 style approach, in which the rating dimensions are not static or predefined.

The goal was to develop and evaluate different recommendation schemes that take item-specific tag preferences into account when generating rating predictions. In addition, we proposed a metric to automatically derive user- and item-specific tag preferences from the overall ratings based on item similarities, in order to demonstrate that quality improvements can be achieved even when the tag preference data is not explicitly given. The results of the evaluation on two data sets show that a measurable accuracy

---

[10]http://www.delicious.com

improvement can be achieved. The limitations of the evaluation are addressed in Chapter 7.

Since our approach of rating items by rating tags introduced in this chapter is different from the existing recommendation approaches based on Social Web tagging data (compare, e.g., with the selection of recent approaches presented in Section 2.3), we see the following opportunities for further improvements and developments.

- **Further experiments and user interfaces.** Further experiments with real tag preference data have to be conducted. A particular question to be answered in that context is that of an appropriate user interface (see also [Vig et al., 2009] or [Vig et al., 2010]) because Web users are currently not acquainted to the interaction pattern "providing ratings for tags". Intuitively, interfaces that allow users to rate tags on a 5-star scale with half-star increments or allow users to classify tags in two or three categories such as "like", "dislike", or "indifferent" seem appropriate. However, we aim to explore different visualizations to stimulate more precise ratings.

- **Combination with tag recommenders.** Different techniques to *tag recommendation* have been developed in the last years to stimulate users to use a more consistent set of tags in the annotation process, see, for example, our LocalRank algorithm presented in Chapter 3. We expect that the value of item-specific tag preferences is even higher, when the overall set of tags used in the data set is more consistent.

- **New tag quality metrics.** We have stated in Section 4.5.1 that our approach of rating items by rating tags calls for a new quality requirement to tags: tags must be appropriate for ratings. Therefore, new tag quality metrics can be developed in order to, e.g., improve the overall performance of recommendation algorithms that are based on tag preferences.

- **Better explanations.** Tags can also be a helpful means to generate explanations for the end user. Explanations for recommendations are one of the current research topics in the recommender systems area because they can significantly influence the way a user perceives the system [McSherry, 2005]. In [Tintarev and Masthoff, 2007a], for example, seven possible advantages of an explanation facility are described. In [Vig et al., 2009], the authors have evaluated explanation interfaces which use tag relevance and tag preference as two key components. Based on appropriately designed explanation interfaces, the different aspects of explanations as discussed in [Tintarev and Masthoff, 2007a] (such as transparency, trust, effectiveness, and satisfaction) can be analyzed in different user studies. Again, also the question of the appropriate end-user visualization has to be answered.

In the subsequent Chapters 5 and 6 we examine the role of tag preferences in helping users understand their recommendations. If tags are both user- *and* item-specific, more personalized and detailed, multi-dimensional explanations can be provided.

We believe that further questions related to tag preferences in the Social Web recommendation process will arise.

Chapter 5

Evaluation of explanation interfaces in the form of tag clouds

Current research has shown the important role of explanation facilities in recommender systems based on the observation that explanations can significantly influence the user-perceived quality of such a system [Tintarev and Masthoff, 2012]. In this chapter, we present and evaluate explanation interfaces in the form of *tag clouds*, which are a frequently used visualization and interaction technique on the Web [Lohmann et al., 2009]. We report the result of a user study in which we compare the performance of two new explanation methods based on personalized and non-personalized tag clouds with a previous explanation approach. Overall, the results show that explanations based on tag clouds are not only well-accepted by the users but can also help to improve the efficiency and effectiveness of the explanation process. Furthermore, we provide first insights on the value of personalizing explanations based on the concept of item-specific tag preferences proposed in Chapter 4.

5.1 Introduction

Recommender systems point the users to possibly interesting or unexpected items, thereby increasing sales or customer satisfaction on modern e-commerce platforms [Linden et al., 2003; Senecal and Nantel, 2004; Zanker et al., 2006; Dias et al., 2008; Hegelich and Jannach, 2009]. However, personalized item lists alone might be of limited value for the end users when they have to decide between different alternatives or when they should assess the quality of the generated recommendations. In other words, only showing the recommended-item lists will make it hard for the users to decide whether they actually trust that the recommended items are actually relevant without inspecting all of them in detail.

The capability of a recommender system to *explain* the underlying reasons for its proposals to the user has increasingly gained in importance over the last years both in academia and industry, see [Tintarev and Masthoff, 2011] for a recent overview. Amazon.com, for example, as one of the world's largest online retailers, allows their online users not only to view the reasons for its recommendations but also to influence the recommendation process and exclude individual past purchases from the recommendation process.

Already early research studies in the area – such as the one by [Herlocker et al., 2000] – have shown that the provision of explanations and transparency of the recommendation process can help to increase the user's acceptance of collaborative filtering recommender system. Later on, [Tintarev and Masthoff, 2011] analyzed in greater detail the various goals that one can try to achieve with the help of an explanation facility. Among other aims, explanations can help the user to make his or her decision more quickly, convince a customer to buy something, or develop trust in the system as a whole.

The question is not only what makes a *good* explanation but also how can we automatically construct explanations which are *understandable* for the online user. With respect to the second aspect, Herlocker et al., for example, experimented with different visual representations such as histograms of the user's neighbors' ratings. Later, Bilgic and Mooney, observed that such neighborhood-style explanations are good at promoting items but make it harder for users to evaluate the true quality of a recommended item

[Bilgic and Mooney, 2005]. Thus, they introduced a different, text-based explanation style ("keyword-style explanations") in order to overcome this problem which can in the long term lead to dissatisfaction with the system.

We believe that the development of high-quality explanations in recommender systems can further profit from considering different views from related research communities such as intelligent systems, human-computer interaction, and information systems. Therefore, we extend the works of [Sen et al., 2009b] and [Kim et al., 2010a] and propose to use the state-of-the-art user interface of *tag clouds* as a means to explain recommendations because tag clouds have become a popular means in recent years to visualize and summarize the main contents, e.g., of a Web page or news article [Halvey and Keane, 2007; Lohmann et al., 2009]. Our hypothesis is that tag clouds are more suitable than keyword-style explanations to achieve the following goals of an explanation capability which are introduced in the next section: user satisfaction, efficiency, and effectiveness [Tintarev and Masthoff, 2007a]. As a whole, by achieving these goals, we aim to increase the users' overall trust in the recommender.

The chapter is organized as follows: In Section 5.2 the quality factors for recommender system explanations are described in detail. Related work is integrated into the discussion of the quality factors. Section 5.3 describes the different explanation interfaces, which we evaluated in a user study. Details of the study as well as the discussion of the results are finally presented in Sections 5.4 and 5.5 respectively.

5.2   Evaluation factors of explanations in recommender systems

The concept of explanation has been widely discussed in the research of intelligent systems, especially in knowledge-based systems. An explanation facility enables a system to provide understandable decision support and an accountable problem solving strategy. Therefore explanation is considered as one of the important and valuable features of knowledge-based systems [Berry and Broadbent, 1987]. In recent years, the concept of explanations has also been studied and adopted in the area of recommender systems [Pu and Chen, 2007; Tintarev and Masthoff, 2008a; Vig et al., 2009; Friedrich and Zanker, 2011; Tintarev and Masthoff, 2012]. An explanation can be considered as a piece of information that is presented in a communication process to serve different goals, such as exposing the reasoning behind a recommendation [Herlocker et al., 2000] or enabling more advanced communication patterns between a selling agent and a buying agent [Jannach et al., 2010].

In different works of Tintarev and Masthoff, the authors identify seven possible aims of explanations [Tintarev and Masthoff, 2007a,b, 2011]: transparency (explaining why a particular recommendation is made), scrutability (allowing interaction between user and system), trust (increasing the user's confidence in the system), effectiveness (helping the user make better decisions), persuasiveness (changing the user's buying behavior), efficiency (reducing the time used to complete a task), and satisfaction (increasing usability and enjoyment). In the literature, these possible aims of explanations are considered as evaluation factors that can be used to measure the quality and value of explanations provided by a recommender system. In this work, we consider the factors efficiency, effectiveness, persuasiveness, transparency, user satisfaction, and trust.

- *Efficiency:* In the context of recommender systems, the efficiency of explanations often plays a role in conversational recommender systems [Tintarev and Masthoff, 2007a]. Efficiency can be defined from two different perspectives. From the user perspective, efficiency means the ability of an explanation to help the user reduce the decision time or the required cognitive effort. Thompson et al., for example, measure efficiency by computing the total interaction time between the user and the recommender system until the user found a suitable item [Thompson et al., 2004]. On the other hand, from the system perspective, efficiency refers to how quickly the system can make the recommendations. David McSherry, for instance, measures efficiency through the number of dialogue steps between the user and the recommender system before the final recommendations are accepted [McSherry, 2005]. For other types of recommender systems efficiency is sometimes calculated by measuring the time used to complete the same task with and without an explanation facility or with different types of explanations. In the user study of [Pu and Chen, 2006], for instance, the authors present two different explanation interfaces to users and compare the time needed to locate a desired item in each interface. The observed differences for the different scenarios are then used as an indicator of system-side efficiency. We will consider efficiency from the user perspective and analyze how long a user needs to make a decision.

- *Effectiveness:* Explanation effectiveness can be defined as the ability of an explanation to help the users to correctly identify the actual quality or suitability of the recommended items [Bilgic and Mooney, 2005; Tintarev and Masthoff, 2012]. This way, users are then able to filter out irrelevant items and make better decisions. Bilgic and Mooney argue that effectively explaining the recommendations is an important aspect of increasing the utility and usability of a recommender system [Bilgic and Mooney, 2005]. One possible approach to measure effectiveness is to examine if the user is satisfied with his or her decision. Vig et al. present four kinds of explanations to users and let users rate how well different explanations help the users decide whether they like a recommended item [Vig et al., 2009]. An explanation which helps user make better decisions is considered effective. In [Bilgic and Mooney, 2005], effectiveness is measured by the closeness between the user's estimate of the quality or appropriateness of an item and the actual quality or utility of the recommended items. The used procedure is as follows. First, users are asked to estimate the quality of a recommended item by considering only the explanation generated by the recommender. Afterwards, users should use or "consume" the item (e.g., watch a movie or analyze the item based on more detailed information) and rate the item again based on their real experiences or the additional knowledge. The closeness between the two ratings can be used to measure effectiveness. We will follow this approach in our experiments and use Bilgic and Mooney's metric, as it measures objective effectiveness and is still used in very recent publications, see, e.g., [Tintarev and Masthoff, 2012].

- *Persuasiveness:* Effectiveness is a factor that is highly related to persuasiveness. Persuasiveness can be defined as the ability of an explanation to convince the user to adopt or disregard certain items and can be inferred from the study of effectiveness [Bilgic and Mooney, 2005]. Strong persuasiveness in explanations may also be used to manipulate the users' opinion to accept some product in a sales promotion, for example. The level of persuasiveness of explanations can be approximated by a measurement which determines to what extent the user's evaluation is changed by the explanations. We propose to divide the concept of persuasiveness into overestimate- and underestimate-oriented persuasiveness. Overestimate-oriented persuasiveness means that the explanations result in the effect that users overestimate the quality of an item. Thus, overestimate-oriented persuasiveness may increase sales through a promotion of the products. On the other hand, underestimate-oriented persuasiveness leads users to underestimate the quality of an item. Such a method could potentially be used to let users underestimate the value of certain products and to direct customers to a certain part of the product spectrum. Therefore, the use of persuasive explanations in recommender systems has to be in line with the business strategy and a proper adoption of persuasive explanations may help to increase sales. However, too much persuasiveness can hurt the user's trust in the long run [Tintarev and Masthoff, 2007b]. Bilgic and Mooney argue that effectiveness is more important than persuasiveness in the long run as greater effectiveness can help to establish trust and attract users [Bilgic and Mooney, 2005]. In our experimental studies, we will consider persuasiveness together with effectiveness.

- *Transparency:* In early knowledge-based systems, system-generated explanations have already been used to make the system more transparent. An increased level of transparency enabled classical knowledge-based systems to produce more credible predictions or more accountable decision support in various domains. Examples can be found in the financial and medical industry [Rowe and Wright, 1993; Ong et al., 1997]. In recent years, the concept of transparency has been increasingly adopted also in the context of explanations of recommender systems, where the goal is to expose the reasoning behind a recommendation [Herlocker et al., 2000]. In recommender system research, most of the presented systems in the past have focused mainly on the input (e.g., preference acquisition and user modeling) and output (e.g., a recommender's prediction accuracy), leading to a "black box" perception among the users. Transparency reveals parts of the internal recommendation process and allows users to understand why certain items are recommended. Transparency in recommender systems is considered as an important factor that contributes to users building trust in the system [Swearingen and Sinha, 2002]. Even if the quality of recommendations is sometimes poor, users might tend to like the trustworthy system since they understand how the recommendations are generated. In [Vig et al., 2009], the user-perceived transparency is called *justification* which differs from objective transparency as it may not reveal the actual mechanisms of the recommender algorithm. However, Vig et al. list several reasons why justifications might be preferred, e.g., the algorithm may be too complex or not intuitive or the algorithm details may be protected. We will measure the user-perceived transparency based on a survey in which the users participate after interacting with the system[1].

- *Trust:* In the context of recommender systems, trust can be seen as a user's willingness to believe in the appropriateness of the recommendations and making use of the recommender system's capabilities [Cramer et al., 2008]. Trust therefore in a sense represents how credible and reliable the system is. As mentioned above, transparency is positively related to trust. In the field of recommender systems, definitions of trust usually fall into two categories: interpersonal trust and system trust [O'Donovan and Smyth, 2005b]. Interpersonal trust means that one user trusts another user, e.g., in a social network. This trust relationship between users can be incorporated into recommender algorithms to build trust-aware recommender systems, which can alleviate the problems such as data sparseness and cold start [Massa and Bhattacharjee, 2004]. The second category is system trust, i.e., user's trust in the recommender system, which can be seen as a long-term relationship between a user and a recommender system. In this sense, trust can be measured by perceived competence and trusting intentions [Pu and Chen, 2006, 2007] or indirectly measured by user loyalty and increased sales [Tintarev and Masthoff, 2007a]. Unlike the other evaluation factors, we will examine trust later on in conjunction with the other factors and not separately since we see trust as an abstract factor which is difficult to measure.

- *Satisfaction:* User satisfaction measures the user's perception of the explanation quality [Tintarev and Masthoff, 2012]. Similar to trust, the user's overall satisfaction with a system is assumed to be strongly related to the perceived recommendation or explanation quality of a system [Cosley et al., 2003]. Previous work such as [Swearingen and Sinha, 2002] or [McCarthy et al., 2004] has demonstrated that using explanations can increase the users' overall satisfaction with a recommender system. One popular method of measuring explanation satisfaction is to directly ask users to which extent they like the explanations (*"How good do you think this explanation is?"*) [Tintarev and Masthoff, 2012]. We will also follow this approach and directly ask users. Furthermore, we will analyze whether other factors (efficiency, effectiveness, and transparency) measurably contribute to user satisfaction.

In this section, we have discussed the different goals of explanations in recommender systems, all of which will be used to evaluate the value of explanations in this study and the subsequent study presented in Chapter 6. More details about how we measured these factors will be provided in Section 5.4.1.

5.3   Overview of the evaluated explanation interfaces

In this section we will provide an overview of the three different explanation interfaces which were evaluated in our first study presented in this chapter: keyword style explanations (KSE), tag clouds (TC), and personalized tag clouds (PTC). KSE, which relies on automatically extracted keywords from item descriptions, is used as the baseline method because this visualization approach has performed best according to effectiveness in previous work [Bilgic and Mooney, 2005]. The new methods TC and PTC, however, make use of user-contributed tags, which are a highly popular means of organizing and retrieving content in the Web 2.0.

Keyword-Style Explanations (KSE)

The KSE interface as shown in Figure 5.1 has performed the best in the study by [Bilgic and Mooney, 2005]. The interface consists of a top-20 list of keywords, which are assumed to be the most important

---

[1]Transparency results are not reported in this study but in the subsequent study presented in Chapter 6.

ones for the user. Note that KSE – in contrast to the other interfaces – does not make use of user-generated tags at all. Instead, it relies on keywords that are automatically extracted from the content description of each item. Internally, an item description has different "slots". Each slot represents a "bag of words", that is, an unordered set of words together with their frequencies. Since we are considering the movie domain in our study, we organize a movie's content description using the following five slots: director, actors, genre, description, and related-titles. We have collected relevant keywords about director, actors, genre, and related-titles from the IMDb Web site and the MovieLens data set[2]. The data for the description slot was collected by crawling movie reviews in Amazon.com as well as synopsis information collected from Amazon.com, Wikipedia, and moviepilot[3].

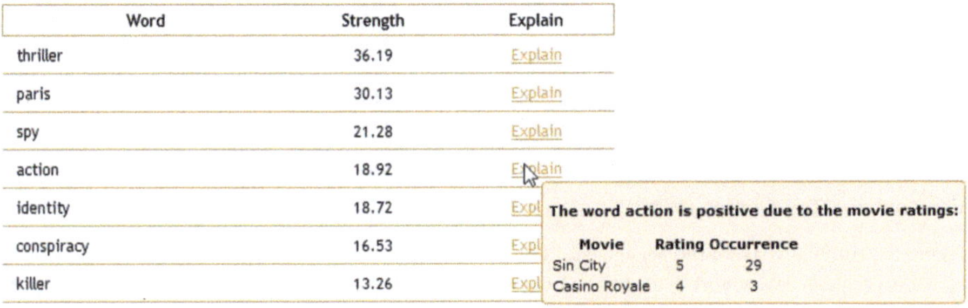

Word	Strength	Explain
thriller	36.19	Explain
paris	30.13	Explain
spy	21.28	Explain
action	18.92	Explain
identity	18.72	Expl
conspiracy	16.53	Expl
killer	13.26	Expl

The word action is positive due to the movie ratings:

Movie	Rating	Occurrence
Sin City	5	29
Casino Royale	4	3

Figure 5.1: Keyword style explanation (KSE).

In the KSE-style approach, the importance of a keyword is calculated using the following formula: $strength(k) = t * userStrength(k)$, where t stands for the number of times the keyword k appears in slot s. The function $userStrength(k)$ expresses the target user's affinity towards a given keyword. This aspect is estimated by measuring the odd ratios for a given user, that is, how much more likely a keyword will appear in a positively rated example than in a negatively rated one. More formally: $P(k|positive\ classification)/P(k|negative\ classification)$. A naïve Bayesian text classifier is used for estimating the probabilities. More details about the KSE-style interface are given in [Bilgic and Mooney, 2005].

Beside the list of important keywords, the KSE explanation interface provides a link ("Explain") for each keyword that opens a pop-up window containing all the movies that the user has rated which contain the respective keyword. In this pop-up window the user is provided with both the user's past rating for the movie and the number of times the keyword appears in the corresponding slot.

Note that in [Bilgic and Mooney, 2005], the KSE approach performed best in the book domain with respect to effectiveness. However, the evaluation of efficiency and satisfaction is not part of their work but will be analyzed in this study.

Tag Clouds (TC)

Tag clouds as shown in Figure 5.2 have become a frequently used visualization and interaction technique on the Web. They are often incorporated in Social Web platforms such as Delicious and Flickr[4] and are used to visually present a set of words or user-generated tags. In such tag clouds, attributes of tags such as font size, weight, or color can be varied to represent relevant properties like relevancy or frequency of a keyword or tag. Additionally, the position of the tags can be varied. Usually, however, the tags in a cloud are sorted alphabetically from the upper left corner to the lower right corner.

In our basic approach of using tag clouds as a not-yet-explored means to explain recommendations, we only varied the font size of the tags, i.e., the larger the font size, the stronger the importance of the tag. We simply used the number of times a tag was attached to a movie as a metric of its importance. The

---

[2]http://www.imdb.com, http://www.grouplens.org/node/73
[3]http://www.amazon.com, http://www.wikipedia.org, http://www.moviepilot.de
[4]http://www.delicious.com, http://www.flickr.com

underlying assumption is that a tag which is often used by the community is well-suited to characterize its main aspects. For all other visual attributes we used the standard settings (font sizes etc.). An analysis of the influence of the different attributes in explanation scenarios can be addressed in future work.

Figure 5.2 (a) shows an example for a movie explanation using the TC interface. Tags such as "Quentin Tarantino" or "violence" have been used by many people and are thus displayed in a larger font size.

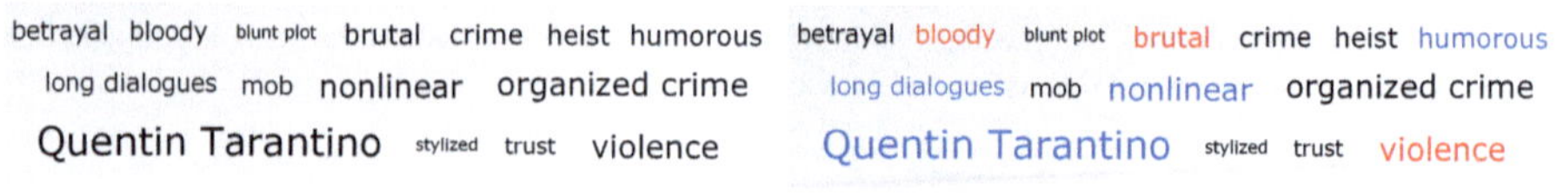

(a) Tag cloud (TC).   (b) Personalized tag cloud (PTC).

Figure 5.2: Tag cloud explanation interfaces.

Personalized Tag Clouds (PTC)

Figure 5.2 (b) finally shows the personalized tag cloud (PTC) interface, which unlike the TC interface is able to exploit the concept of item-specific tag preferences [Gedikli and Jannach, 2010c; Vig et al., 2010] introduced in Chapter 4. The idea of tag preferences is that users should be allowed to assign preferences to tags in order to express their feelings about the recommendable items in more detail. Thus users are not limited to the one single overall vote anymore. In the movie domain, tag preferences can give us valuable information about what users particularly liked/disliked about a certain movie, e.g., the actors or the plot. The PTC interface represents a first attempt to exploit such tag preferences for explanation purposes.

In contrast to the TC interface, we vary the color of the tags according to the user's preference attached to the tag. Blue-colored tags are used to highlight aspects of the movie toward which the user has a positive feeling, whereas tags with a negative connotation are shown in red. Neutral tags, for which no particular preference is known, are shown in black. Again, the font size is used to visualize the importance or quality of a tag. An example of the PTC interface for a crime movie is shown in Figure 5.2 (b). According to the explanation, the user is assumed to like this movie because of its director *Quentin Tarantino*, whereas *violence* and *brutality* are reasons not to watch this movie.

As explanations are usually presented for items which the user does not know yet, we have to first *predict* the user's feeling about the tags attached to a movie. For this purpose, we analyze the tag preference distribution of the target user's nearest neighbors and decide whether the target user will like, dislike, or feel neutral about the item features represented by these tags. In order to predict a preference for a particular tag, the neighbors preferences for this tag are summed up and normalized to our preference scale for tags. Note that in our study users were able to give preferences to tags on a 5-point scale with half-point increments (0.5 to 5). If the predicted preference lies between $[0.5, 2.0]$ or $[3.5, 5.0]$, we will assume negative or positive connotation respectively; otherwise we will assume that the user feels neutral about the tag.

It is important to know that the interfaces KSE and PTC are personalized, whereas TC represents a non-personalized explanation interface.

5.4   Experimental setup

5.4.1   Experimental procedure

We have conducted a within-subjects user study (see Section 2.2.2) in which each subject was confronted with all explanation interfaces presented above. In this section, we will shortly review the experimental setup, which consisted of two phases.

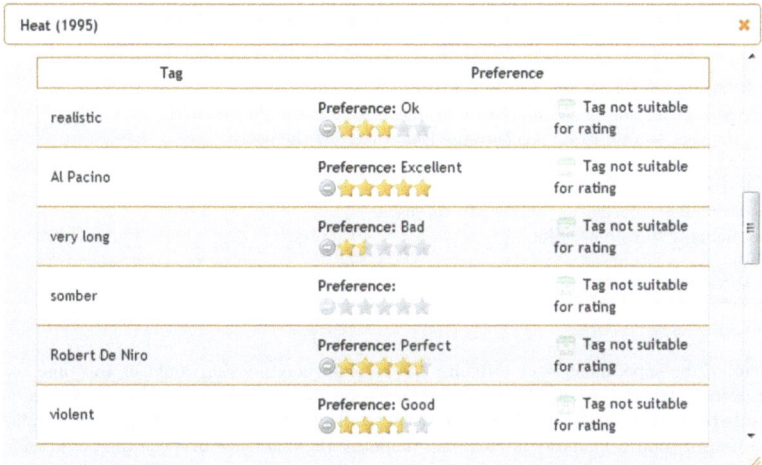

Figure 5.3: Rating (tags of) the movie *Heat (1995)* on a Likert scale of 0.5 to 5.

Experiment - phase 1

In the first phase of the experiment, the participants were asked to provide preference information about movies and tags to build the user profiles. The users had to rate at least 15 out of 100 movies[5]. After rating a movie, a screen was shown (Figure 5.3) in which users could rate up to 15 tags assigned to the movie[6]. On this screen, users could rate an arbitrary number of tags; skip tags, in case they thought that they were not suitable for a given movie; or explicitly mark tags as inappropriate for rating. Note that users were not allowed to apply their own tags as we want to ensure that we have a reasonable overlap in the used tags.

Experiment - phase 2

In the second phase, which took place a few weeks after the first session, the subjects used a recommender system[7] which presented them movie recommendations based on the user profile from the first phase. In addition, the different explanation interfaces were shown to the user. In the following, we will introduce our experimental procedure summarized in Procedure 1, which extends the procedure proposed by Bilgic and Mooney [Bilgic and Mooney, 2005].

The evaluation system randomly selected a tuple (r, e) of possible recommendation and explanation pairs and presented the movie recommendation r using explanation interface e to the end-user without showing the title of the movie. The user was then expected to provide a rating for the movie by solely relying on the information given in the explanation (lines 1-6). The selection order is randomized to minimize the effects of seeing recommendations or interfaces in a special order [Bilgic and Mooney, 2005]. If the users recognized a movie based on the information presented in an explanation, they could inform the system about that. No rating for this movie/interface combination was taken into account in this case to avoid biasing effects. We additionally measured the time needed by the users to submit a rating as to measure the *efficiency* of the user interface. These steps were repeated for all movie/interface combinations. Afterwards, we again presented the recommendations to the user, this time showing the complete movie title and links to the corresponding movie information pages at Wikipedia, Amazon.com, and IMDb. We provided information about movies to reduce the time needed

---

[5]We have limited the number of movies to 100 in order to be able to find nearest neighbors in the PTC approach.
[6]The tags were taken from the "Movie-Lens 10M Ratings, 100k Tags" data set (http://www.grouplens.org/node/73).
[7]We used a classical user-based collaborative filtering algorithm.

Procedure 1 Experimental procedure

1: **R** = Set of recommendations for the user.
2: **E** = Set of explanation interfaces KSE, TC, PTC.
3: **for all** randomly chosen (r, e) in **R** x **E** **do**
4:     Present explanation using interface e for recommendation r to the user.
5:     Ask the user to rate r and measure the time taken by the user.
6: **end for**
7: **for all** recommendation r in **R** **do**
8:     Show detailed information about r to the user.
9:     Ask the user to rate r again.
10: **end for**
11: Ask the user to rate the explanation interfaces.

---

for completing the experiment since watching the recommended movies would be too time consuming. The users were instructed to read the detailed information about the recommended movies and were then asked to rate the movies again (lines 7-10). According to [Bilgic and Mooney, 2005], from the point of view of an end-user, a good explanation system can minimize the difference between ratings provided in the lines 5 (explanation rating) and 9 (actual rating). Thus we can also measure *effectiveness/persuasiveness* by calculating the rating differences. At the end of the experiment, the users were asked to give feedback on the different explanation interfaces (as to measure *satisfaction* with the system) by rating the system as a whole on a 0.5 (lowest) to 5 (highest) rating scale (line 11). Again, the order was randomized to account for biasing effects.

5.4.2 Participants of the study

We recruited 19 participants from five different countries. Most of them were students at our institution with their age ranging from 22 to 37 (average age was 28 years). Ten participants declared high interest in movies, whereas eight were only to a certain extent interested in movies. One person was not interested in movies at all. Table 5.1 shows some of the demographic characteristics of the participants.

Gender	Female	4 (21.05%)
	Male	15 (78.95%)
Education	A-level	9 (47.37%)
	Bachelor	1 (05.26%)
	Master	7 (36.84%)
	PhD	2 (10.53%)
Nationality	5 countries	(Germany, Turkey, Brazil, etc.)
Age	21-25	5 (26.32%)
	26-30	10 (52.63%)
	31-40	4 (21.05%)
Interest in movies	High	10 (52.63%)
	Normal	8 (42.11%)
	Low	1 (05.26%)

Table 5.1: Demographic characteristics of participants (total 19).

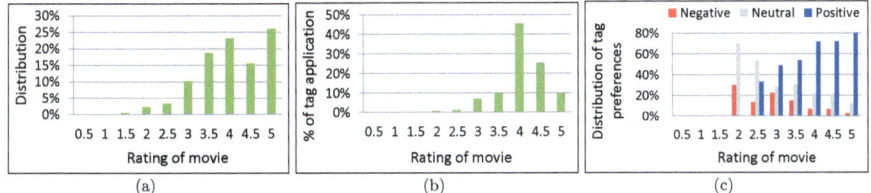

Figure 5.4: Distribution of (a) movie ratings, (b) tag applications over movie ratings, and (c) negative, neutral, and positive tags applied to movies with different ratings.

5.4.3 Data collected in the experiment

The participants provided a total of 353 overall movie ratings and 5,295 tag preferences. On average, each user provided 19 movie ratings and 279 tag preferences and assigned 15 tag preferences to each rated movie. Because participants were also allowed to repeat phase 2 of our user study, we collected a total of 848 explanation ratings (on average 45 ratings per user).

Figure 5.4 (a) shows the distribution of the movie ratings collected in our study. It can be seen that users preferred to rate movies they liked, i.e., a *positivity bias* is present among the participants which is in line with the findings of other researchers [Marlin et al., 2007; Vig et al., 2010]. Vig et al. show that the positivity bias is also present for the taggers, that is, taggers apply more tags to movies they liked compared to movies they rated badly [Vig et al., 2010]. This finding is also consistent with our results, as shown in Figure 5.4 (b). Users applied four times more tags to movies they rated with 4 or higher compared to movies to which they gave less than 4 points. Figure 5.4 (b) shows another interesting effect, which is only partly visible in the data of [Vig et al., 2010]. Users applied seven times more tags to movies rated with 4 or 4.5 points compared to movies rated with 5 points – the highest rating value – although there are more movies rated with 5 points than with 4 or 4.5 points, as shown in Figure 5.4 (a). We believe that this effect may be due to users' demand for justifying non-5-point ratings, i.e., users want to explain to the community *why*, in their opinion, a particular movie does not deserve a 5 point rating.

Figure 5.4 (c) finally shows the distribution of negative, neutral, and positive tags applied to movies with different ratings. For clarity reasons, we have classified the tag preferences into the tag preference groups *negative* (< 2.5 points), *neutral* ($2.5 - 3$ points), and *positive* (> 3 points). As expected, a user's movie rating has a strong influence on the tag preferences assigned to a movie. The number of positive (negative) tag preferences increases (decreases) with the overall movie rating. Again, the results are comparable with those reported in [Vig et al., 2010].

5.5 Hypotheses, results, and discussion

We tested three hypotheses. First, we hypothesized that the tag cloud interfaces TC and PTC enable users to make decisions faster (**H1:Efficiency**). We believe this as we think the visual nature of a tag cloud allows users to grasp the content information inside a cloud faster compared to KSE, which are organized in a more complex tabular structure. We also believe that users enjoy explanations from TC and PTC more than in the KSE style as we assume that tag cloud explanations are easier to interpret for the end user (**H2:Satisfaction**). We further conjecture that users make better buying decisions when their decision is based on TC or PTC rather than KSE (**H3:Effectiveness**). We believe this because we think that compared to TC or PTC, there is a higher risk of misinterpreting KSE because users always have to consider both the keyword and its corresponding numerical importance value, whereas in TC and PTC the importance is encoded in the font size of a tag.

In the following we will have a closer look at the results which are summarized in Table 5.2. We have used the Friedman test with the corresponding post-hoc Nemenyi test in this study as suggested by Demšar for a comparison of more than two systems [Demšar, 2006].

		KSE	TC	PTC	N	α
(a)	Mean time [sec]	30.72	**13.53**	**10.66**	60	0.05
	Standard deviation	19.72	8.52	5.44		
(b)	Mean interface rating	1.87	**3.74**	**3.87**	19	0.05
	Standard deviation	0.90	0.65	0.62		
(c)	Mean difference	-0.46	**-0.13**	**-0.08**	283	0.05
	Standard deviation	1.00	1.01	1.03		
	Pearson correlation	0.54	0.79	0.83		

Table 5.2: (a) Mean time for submitting a rating. (b) Mean response of the users to each explanation interface. (c) Mean difference of explanation ratings and actual ratings. Bold figures indicate numbers that are significantly different from the base cases (N is the sample size and α is the significance level).

5.5.1 Efficiency

To test our hypothesis of improved efficiency of tag clouds, we analyzed the time measurement data which was automatically collected in our study[8]. Table 5.2 (a) shows the mean times (in seconds) for submitting a rating after seeing the corresponding explanation interface. We can see that the time needed when using the tag cloud approaches is significantly shorter than for KSE. Thus, we can conclude that the data supports hypothesis H1. The data also indicates that the PTC method helps users to make decisions slightly faster than the TC approach, but the difference was not statistically significant.

5.5.2 Satisfaction

Table 5.2 (b) shows the mean response on overall satisfaction of 19 users to each explanation interface based on a Likert scale of 0.5 to 5. It can be seen that users prefer the PTC approach over the TC presentation style and the TC style over the KSE method, which supports hypothesis H2. Again, the differences between the keyword-style explanations and the tag cloud interfaces are significant but no significant difference among the tag cloud interfaces could be found although the data indicates that users favor PTC-style explanations. One possible reason is that tag clouds are in general capable of visualizing the context in a concise manner and can thus help users reduce the time needed to understand the context which in turn increases user satisfaction.

5.5.3 Effectiveness and persuasiveness

Bilgic and Mooney propose to measure effectiveness by calculating the rating differences between explanation rating and actual rating [Bilgic and Mooney, 2005]. If the difference is 0, the explanation and the actual rating will match perfectly, i.e., the explanation helps the user to accurately predict the quality of an item. Otherwise, if the difference is positive (negative), users will overestimate (underestimate) the quality of an item. In this context we talk about the persuasive power of an explanation system.

Table 5.2 (c) shows the mean difference of explanation ratings and actual ratings. The histograms showing the mean differences are presented in Figure 5.5.

The mean differences of the tag cloud interfaces are close to 0 which is an indication that the interfaces are valuable for users to accurately estimate the quality of an item. Note that we have also considered the Pearson correlation between explanation and actual ratings to account for averaging effects. From the user's point of view, a good explanation interface has a mean difference value of 0, a low standard deviation, and a high correlation between both rating values [Bilgic and Mooney, 2005].

Users can estimate item quality most precisely with the help of the PTC interface. TC explanations are also a good estimator for item quality. The KSE interface has a significantly different value of −0.46 which means that KSE cause the user to underestimate the actual rating on average by −0.46. On a 5-point scale with half-point increments an underestimation of −0.46 on average can be considered as important. Note that in [Bilgic and Mooney, 2005], KSE reached a value of 0. We believe that the

---

[8]The experiment was conducted over the Internet. Therefore, timings may be unreliable and may impact efficiency results.

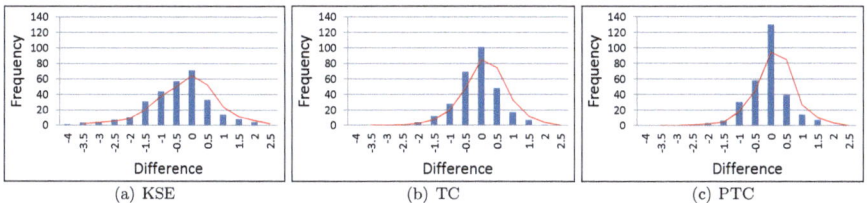

Figure 5.5: Histograms showing the differences between interface and actual ratings.

difference in the mean values comes from the different domains considered in our studies (movie domain vs. book domain). Overall the results support our last hypothesis H3.

Next we will discuss about the tradeoff between effectiveness and persuasiveness and the influence of persuasiveness on the user's trust in a recommender system.

5.5.4 Implications for trust

As mentioned above, effectiveness can be measured by the rating difference before and after the consumption or inspection of a recommended item. Smaller differences are indicators of higher effectiveness. Therefore, if the rating for an item based only on the explanation is the same as the rating after the user has consumed the item, we can consider the explanation as highly effective. In the other case, the limited effectiveness will negatively impact on user satisfaction and the trust in the recommender system.

Consider the following case. A user rates an item with 4 (good) based only on the explanation. After consuming this item, however, the user rates the item with 2 (bad). This means that the user found this item is not as good as expected given only the explanation. In this scenario the user may consider the explanation to be not trustful. We call this effect *positive persuasiveness*, as the system successfully persuades the user to consume/buy the item. Conversely, if the user initially rates the item first with 2 and finally with 4, this means that the explanation does not correctly reflect the truth. In this case, the user may find the explanation to be inaccurate and lose the interest in using this system. We call this effect *negative persuasiveness*. Both positive and negative persuasiveness can cause the loss of trust to users.

The question remains, which form of persuasiveness is better. From a user's perspective, positive persuasiveness may leave the user with the impression that the system is cheating because the system overstates the advantages of the item. This may cause the user to completely abandon the system. However, from a business perspective, if a firm intents to promote a new product or convince the user to adapt a new version of a product, positive persuasiveness may help to increase effects of advertisement and user's familiarity to this product. Negative persuasiveness, on the other hand, has a different effect and may cause the user to suppose that the system does not really take his or her preferences into account. Tintarev and Masthoff showed that users perceived positive persuasiveness (overestimation) to be less helpful than negative persuasiveness (underestimation) [Tintarev and Masthoff, 2008b]. Therefore, we assume it to be a rather "safe" strategy, if we are able to keep the negative persuasiveness level within a certain range.

Overall, we argue that it is important to choose the direction of the persuasiveness according to different cases and goals. We can either align positive persuasiveness with the business strategy or control the negative persuasiveness at an acceptable level.

5.6 Summary

In this chapter, we have presented the results of our first user study in which three explanation approaches were evaluated. We have compared keyword-style explanations, which performed best according to effectiveness in previous work, with two new explanation methods based on personalized and non-personalized tag clouds. A tag cloud is a frequently used visualization and interaction technique which is well accepted and understood by users. The personalized tag cloud interface additionally makes use of the recent idea of item-specific tag preferences which was introduced in Chapter 4. We have evaluated the interfaces on the quality dimensions efficiency, satisfaction, and effectiveness (persuasiveness) and discussed their impact on the user's trust in a recommender system.

The results show that users can make better decisions faster when using the tag cloud interfaces rather than the keyword-style explanations. In addition, users generally favored the tag cloud interfaces over keyword-style explanations. This is an interesting observation because users preferred even the non-personalized explanation interface TC over the personalized KSE interface. We assume that there are factors other than personalization such as the graphical representation, which play a crucial role for effective explanation interfaces. The results also indicate that users preferred PTC over TC. We believe that with PTC users need less time to come to an even better conclusion because the font color of a tag already visualizes a user's feeling about the tag and reduces the risk of misinterpreting a tag. For example, consider the case where users see the tags *Bruce Willis* and *romantic movie* in a tag cloud and wonder whether they will like the performance of their action hero in a romantic movie. We believe that higher user satisfaction, efficiency, and effectiveness have positive impact on the users' overall trust in the recommender system which ensures user loyalty and long term wins.

Although we view content and the visualization to be tightly interrelated in explanations (as done in previous work), experiments can be planned in which the effects of content and visualization are evaluated separately.

The small number of participants is one of the limitations of this user study which can be attributed to the fact that two sessions were needed to finish the experiment. However, the fact that we have applied appropriate statistical tests compensates for this limitation and strengthens the reliability of the results presented in this chapter. Further limitations are discussed in Chapter 7.

In the next chapter we present the results of a broader study on the effects of different explanations styles which involves more participants. In particular, our study includes the evaluation of further quality dimensions such as transparency; in addition, we estimate a user's tag ratings automatically in order to reduce the time needed for completing the experiment.

Chapter 6

An analysis of the effects of using different explanation styles

When explaining recommendations to the customers, one of the main challenges is how to select the appropriate presentation interface for explanations. Good explanations not only have to be easily understandable by the end user but should also help the user to make good decisions [Bilgic and Mooney, 2005; Tintarev and Masthoff, 2012]. This chapter addresses the question of how explanations can be communicated to the user in the best possible way. To that purpose, we analyze ten different explanation interfaces with respect to six evaluation factors in a user study. The explanation interfaces used in the study include both known visualizations from the literature as well as our newly proposed interfaces based on tag clouds introduced in the last chapter. Our study reveals that tag cloud explanations are particularly helpful to make recommendations transparent to the user and to finally increase user satisfaction even though they demand higher cognitive effort from the user. Based on these insights and in-depth observations, we derive a set of possible guidelines for designing or selecting the most suitable explanation interface for a recommender system.

6.1 Introduction

One possible approach to support the end user in the decision making process and to increase the trust in the system is to provide an explanation for why a certain item has been recommended [Herlocker et al., 2000; Bilgic and Mooney, 2005; Pu and Chen, 2006; Tintarev and Masthoff, 2007a,b; Friedrich and Zanker, 2011]. There are many approaches of explaining recommendations – even non-personalized ones. An example of a non-personalized explanation would be Amazon.com's *"Customers who bought this item also bought..."* label for a recommendation list, which also carries explanatory information.

This chapter deals with questions of how explanations should be communicated to the user in the best possible way. This includes both questions of the *visual representation* as well as questions of the *content* to be displayed. In general, the type and depth of explanations a recommender system can actually provide depend on the types of knowledge and/or algorithms that are used to generate the recommendation lists. In knowledge-based recommendation or advisory approaches, explanations can be based on the rule base which encodes an expert's domain knowledge and the explicitly acquired user preferences [Felfernig et al., 2007]. For the most prominent type of recommender systems, that is, collaborative filtering recommenders, [Herlocker et al., 2000] and [Bilgic and Mooney, 2005] have proposed various ways of explaining recommendations to the user. Herlocker et al. have also shown that explanation interfaces can help to improve the overall acceptance of a recommender system.

In this chapter, we continue the line of work of [Herlocker et al., 2000], [Bilgic and Mooney, 2005], [Tintarev and Masthoff, 2007a], [Vig et al., 2009], [Tintarev and Masthoff, 2012], and our work presented in the last chapter and provide the following contributions.

1. We present a more in-depth analysis of our newly proposed explanation interfaces from Chapter 5 which use tag clouds as a means for visualization and which are based on detailed tag preferences

of users [Vig et al., 2010; Gedikli and Jannach, 2010c, 2013].

2. Acquiring explicit tag preferences in the sense of [Vig et al., 2010] is costly and can be cumbersome for the user. Therefore, as an extension to our previous study in Chapter 5, we present results for tag cloud explanations which are based on tag preferences that are automatically derived from the item's overall ratings with a new technique. Thus, we avoid acquiring them interactively, which we see as an important step toward the practical applicability of explanations based on tag preferences.

3. In the existing literature on recommender system explanations, authors often focus their analysis on a certain number of explanation goals [Herlocker et al., 2000; Bilgic and Mooney, 2005; Pu and Chen, 2006, 2007; Symeonidis et al., 2009] or explanation types [Vig et al., 2009; Gedikli et al., 2011b]. In our work, we aim at evaluating different explanation interfaces in a comprehensive manner and consider the desired effects and quality dimensions *efficiency, effectiveness, persuasiveness, transparency, satisfaction*, and *trust* [Tintarev and Masthoff, 2011] in parallel. To that purpose, we conducted a user study involving 105 subjects in which we compare several existing explanation interfaces from the literature ([Herlocker et al., 2000]) with our new ones.

4. We analyze the dependencies between the different effects of explanation interfaces and derive a first set of possible guidelines for the design of effective, transparent, and trustful explanation interfaces for recommender systems and sketch potential implications of choosing one or the other interface.

5. We go beyond a theoretical discussion of the interdependencies between more than two quality dimensions (as was done in the last chapter) and conduct an experiment where we analyze the influence of efficiency, effectiveness, and transparency on user satisfaction. Most of the related work only focuses on one explanation goal or analyzes the trade-off relation between two quality dimensions (see Table 2 in [Tintarev and Masthoff, 2012]). Therefore, we see this work as a first step in this direction.

The chapter is organized as follows: Section 6.2 introduces the different explanation interfaces compared in our study. In Section 6.3 the experimental setup is described in more detail. Section 6.4 provides a discussion of the obtained results and contains our first set of possible design guidelines. Section 6.5 finally summarizes the main findings of this work and gives an outlook on future work.

6.2   Overview of the evaluated explanation interfaces

In this section, we will give an overview of the interfaces which were evaluated in our study with respect to the above-mentioned quality dimensions efficiency, effectiveness, persuasiveness, transparency, satisfaction, and trust; see Section 5.2 for a detailed description of the various quality dimensions. For better readability, we will only discuss and depict a selection of interfaces here; the remaining ones are described in Appendix B.1.

In total, we compared ten different explanation interfaces. Seven of them were proposed in the existing literature by [Herlocker et al., 2000]; two interfaces are using tag clouds and have been introduced in the last chapter; the final interface is based on pie charts and represents a different visualization of a particular explanation interface proposed by [Herlocker et al., 2000].

Table 6.1 summarizes the ten interfaces evaluated in our study. For convenience, we will make use of the short interface names in the rest of the paper. We define a personalized explanation interface as one that depends on the target user, whereas a non-personalized interface only depends on the target item. The "Content data" field in addition indicates if an interface makes use of domain specific *content* data. We define content data as manually created or automatically extracted item descriptions such as a movie's plot keywords. It is important to know that in this study we view content and the visualization to be tightly interrelated in explanations (as done in previous work [Herlocker et al., 2000; Bilgic and Mooney, 2005; Vig et al., 2009; Gedikli et al., 2011b]) and do not evaluate effects of content and visualization separately. We will discuss individual aspects in particular of our newly proposed methods in the following sections.

Rank	Interface long name	Interface short name	Person- alized	Content data
1	Histogram with grouping	`clusteredbarchart`	yes	no
3	Neighbor ratings histogram	`barchart`	yes	no
4	Table of neighbors rating	`neighborsrating`	yes	no
7	MovieLens percent confidence in prediction	`confidence`	yes	no
10	# neighbors	`neighborscount`	yes	no
15	Overall percent rated 4+	`rated4+`	no	no
21	Overall average rating	`average`	no	no
	Tag cloud	`tagcloud`	no	yes
	Personalized tag cloud	`perstagcloud`	yes	yes
	Pie chart	`piechart`	yes	no

Table 6.1: Explanation interfaces evaluated in our study, along with their accompanying ranking according to [Herlocker et al., 2000].

6.2.1 Herlocker et al.'s explanation methods

In [Herlocker et al., 2000], twenty-one different explanation interfaces were compared. In their study, they considered *persuasiveness* as the only quality dimension and determined a ranked list of the best-performing interfaces in this respect. The "Rank" column in Table 6.1 contains this ranking of interfaces as observed in [Herlocker et al., 2000]. After applying significance tests, they organized the list in three different groups, where the observed differences within the interfaces in each group were not statistically significant.

Since we aim to analyze explanation interfaces in several dimensions, we did not simply select the best-performing interfaces from their study but picked interfaces from all three groups. The reason is that interfaces which perform well in one dimension can perform poorly in other dimensions and vice versa[1]. In our study, we aimed to cover a wide range of existing interface types and popular ones of today. However, in order to keep the experiments manageable, we had to limit ourselves to a spectrum of interfaces which covers a variety of different types. We selected the interfaces from Herlocker et al.'s study as follows.

- From the top of their list we included the interfaces called *histogram with grouping*, which performed best in their study (see Figure 6.1), the *neighbor ratings histogram* (ranked 3rd), and *table of neighbors rating* (ranked 4th). As mentioned above, we also included a pie chart based interface which represents a pie chart visualization of the same data presented in the *table of neighbors rating* interface.

- From the middle block of their list we selected the interfaces *MovieLens percent confidence in prediction*, *# neighbors*, and *overall percent rated 4+*, which were on ranks 7, 10, and 15 respectively.

- From the end of their list, we picked the interface *overall average rating* (ranked on place 21), which performed worst in their study according to the quality dimension persuasiveness. We decided to include the interface due to its popularity on large-scale Web sites, see Figure 6.2.

Regarding the specific selection of interface types from Herlocker et al.'s study, published in 2000, our hypothesis is that the value of explanations depends on different quality factors[2] and that the perception of the interfaces changes over time, which is also supported by our observations reported later on in Section 6.4.

---

[1] An example of such a trade-off (effectiveness vs. persuasiveness) was discussed in the last chapter. Another example for a trade-off (effectiveness vs. satisfaction) can be found in [Tintarev and Masthoff, 2012].

[2] In Herlocker et al.'s paper, the interfaces were only analyzed from a promotion perspective (persuasiveness).

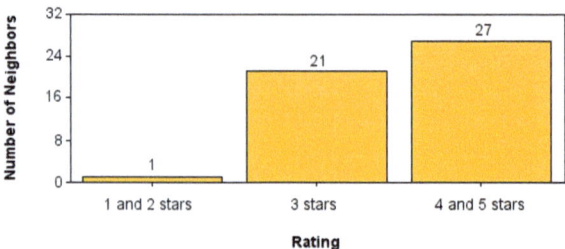

Figure 6.1: The *histogram with grouping* interface, which performed best in the study of [Herlocker et al., 2000]. The X-axis shows the different rating categories (low, medium, high); on the Y-axis, the number of neighbors who gave a certain rating is given.

User rating: ⭐⭐⭐⭐⭐⭐⭐⭐ 8.2/10 144,273 ratings »
Top 250 #166 (Rate now!)

Figure 6.2: IMDb's popular *overall average rating* interface, which performed worst in the study of [Herlocker et al., 2000].

6.2.2 Tag cloud based explanation interfaces

In Chapter 5, we introduced tag clouds as another possible way to visualize explanations in recommender systems (see Figure 6.3). Similar to the approach in [Vig et al., 2009], the basic assumption is that each recommendable item can be characterized by a set of tags or keywords which are provided by the user community (e.g., of a movie portal) or which are automatically extracted from some external source.

Non-personalized tag clouds

The non-personalized tag cloud based explanation interface (referred to as `tagcloud`) corresponds to the TC interface presented in the last chapter. Larger font sizes in the tag cloud correlate with a higher relevance of a term which was calculated by counting the number of times a tag was attached to a resource.

In Figure 6.3 (a) we give an example of a movie explanation using a tag cloud. Tags such as "alfred hitchcock" or "serial killer" have been attached to the movie many times and are correspondingly displayed in a larger font size. In contrast to the personalized approach described in the next section, every movie has one single tag cloud and every user of the system can see the same tag cloud as an explanation.

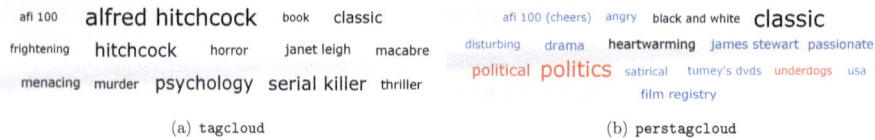

(a) `tagcloud`   (b) `perstagcloud`

Figure 6.3: Tag cloud explanation interfaces.

Personalized tag clouds

Our approach of using personalized tag clouds (`perstagcloud`) is based on the concept of item-specific tag preferences [Gedikli and Jannach, 2010c; Vig et al., 2010], which was introduced in Chapter 4. Personalized tag clouds as shown in Figure 6.3 (b) exploit this more detailed knowledge about a user's preferences to generate individualized tag clouds. The `perstagcloud` interface extends the above-mentioned `tagcloud` and introduces different colors for the elements in the cloud which express the user's sentiment toward a tag. In the example tag cloud in Figure 6.3 (b), we use blue as a color for tags, for which we know or assume that the user has positive feelings about (e.g., the tag "drama"). For red tags such as "politics" and "underdogs", we believe that they represent aspects that the user will not like. Tags for which we do not know the user's sentiment or which are marked as neutral are printed in black.

Compared to the PTC interface from Chapter 5, we further improved the interface `perstagcloud` to make the rationale more transparent for the user. When the user inspects the tag cloud and moves the mouse over a certain tag, we show an additional window which contains all the movies tagged with the respective tag which influenced the tag preference value. The idea was inspired by the keyword-style explanation (KSE)[3] from [Bilgic and Mooney, 2005].

Automatic inference of tag preferences

The generation of personalized tag clouds requires the availability of knowledge about the users' tag preferences. Unfortunately, we are not aware of any publicly available data set that contains such rating information for user provided tags. The "MovieLens 10M Ratings, 100k Tags" data set, for example, which we used in our experiment (Section 6.3.2) does not provide any preference data for the available tags. In our last study on the use of tag cloud explanations, we therefore asked the study participants in a pre-stage of the experiment to explicitly provide their preference information about movies and tags to build the user profile. The effort for the users of such a system is comparably high. They have to provide both an overall rating, optionally attach additional tags in case they do not want to annotate the already existing ones, and finally mark individual tags as positive or negative. Consequently, the number of study participants and the amount of the available rating data were comparably small.

In order to overcome this limitation, we propose to automatically derive an *estimate* of the users' tag preferences if no such information is available. A combination of explicit tag preferences and estimated ones is also possible. For the study presented in this chapter this finally means that no comprehensive profile-building phase as described in the last chapter is required and the experiment can be conducted in one single session.

In order to estimate missing tag ratings from the available overall ratings, we propose to use the function $\hat{r}_{u,i,t}$ shown in Equation (6.1), which we also used in Chapter 4 to generate more accurate predictions based on tag preferences.

$$\hat{r}_{u,i,t} = \frac{\sum_{m \in similarItems(i,I_t,k)} w(m,t) * w_{rpa}(r_{u,m}) * \mathcal{R}(r_{u,m})}{\sum_{m \in similarItems(i,I_t,k)} w(m,t) * w_{rpa}(r_{u,m})} \qquad (6.1)$$

The function is user- and item-dependent and returns an estimated preference value for a tag t given a user-item pair (u,i). The main idea is to consider both the overall ratings of items which are similar to the target item i as well as the relative importance of individual tags in the calculation. The function $similarItems(i,I_t,k)$ returns the collection k of the most similar items to i from I_t, which is the set of all items tagged with t. We use the adjusted cosine similarity metric to calculate the similarity of items. The weight $w(m,t)$ represents the relevance of a tag t for an item m. We use the following simple counting metric to determine the relevance of a tag. Our counting metric gives more weight to tags that have been used by the community more often to characterize the item:

$$w(m,t) = \frac{number\ of\ times\ tag\ t\ was\ applied\ to\ item\ m}{overall\ number\ of\ tags\ applied\ to\ item\ m} \qquad (6.2)$$

When relying on the user's explicit overall rating $r_{u,m}$, no prediction can be made for a tag preference if user u did not rate any item m tagged with t, i.e., if $I_t \cap ratedItems(u) = \emptyset$. We therefore apply the recursive prediction strategy as described in [Zhang and Pu, 2007] and first calculate a prediction

---

[3]Note that we did not include the KSE interface into the comparison pool of this study because no keywords for the considered movies were available and the automatically extracted keywords were of low value. Furthermore, the results presented in the last chapter showed that tag clouds are better accepted by end users than the keyword-style visualization.

for $r_{u,m}$, in case this rating is not available. The function $\mathcal{R}(r_{u,m})$ either returns $r_{u,m}$ if such a rating exists or the estimated value $\hat{r}_{u,m}$. An additional weight value $w_{rpa}(r_{u,m})$ is applied to the recursively predicted value where $w_{rpa}(r_{u,m})$ is defined as follows:

$$w_{rpa}(r_{u,m}) = \begin{cases} 1, & r_{u,m} \text{ is given} \\ \lambda & r_{u,m} \text{ is not given} \end{cases} \qquad (6.3)$$

The combination weight threshold λ is a value between $[0, 1]$. In our study, the parameter λ was set at 0.5 as suggested in [Zhang and Pu, 2007] as an optimal value. We empirically determined $k = 50$ as a suitable value for the neighborhood-size k in Equation (6.1).

In order to classify tags as positive, negative, and neutral, we used the user's average rating value $\overline{r_u}$ to divide the existing tags into two lists, where one list contains tags whose estimated tag preference is above the user's average and another that contains those which are rated below the average. The tags in each list are then sorted by their predicted preference value in ascending order. We then classify the tags in the lower quartile $Q_{.25}$ of the first list as negative and in the upper quartile $Q_{.75}$ of the second list as positive. All the other tags are classified as neutral. This classification is finally used when generating the personalized tag clouds – positive tags are printed in blue, negative tags are printed in red.

6.3 Experimental setup

Next, we will describe the experimental setup used in our user study on the effects of different explanation interfaces.

6.3.1 Experimental procedure

The procedure which we used for evaluating the different explanation interfaces and which is based on the evaluation protocol proposed in [Bilgic and Mooney, 2005] is shown in Procedure 2. Compared to our previous study the whole experiment can be conducted in one single session using Procedure 2 alone.

Procedure 2 Experimental procedure

1: Get sample ratings from the user.
2: **R** = Set of recommendations for the user.
3: **E** = Set of explanation interfaces.
4: **for all** randomly chosen (r, e) in **R** x **E** **do**
5:    Present explanation using interface e for recommendation r to the user.
6:    Ask the user to rate r and measure the time taken by the user.
7: **end for**
8: **for all** recommendation r in **R** **do**
9:    Show detailed information about r to the user.
10:    Ask the user to rate r again.
11: **end for**
12: Ask the user to rate the explanation interfaces.

At the beginning of the study, the participants were asked to provide overall ratings for at least 15 movies from a collection of about 1,000 movies (Procedure 2, line 1). The personalized movie recommendations **R** were computed using a traditional user-based nearest-neighbor collaborative filtering algorithm with a neighborhood-size of 50 (line 2). Note that the recommendation quality for each user can vary. Since the recommendation algorithm was however the same for all users and the used data set contains a number of popular movies[4], we believe that this limitation could be kept under control.

In order to measure the effectiveness of the explanations, we used the metric of [Bilgic and Mooney, 2005] as described in Section 5.2. The evaluation system therefore randomly selected tuples (r, e) of

---
[4]The data set which is described in the next section only contains movies with at least 100 user ratings.

possible recommendation and explanation pairs and presented the movie recommendation r using explanation interface e to the participant. The title of the movie was hidden in order to prevent a bias in the rating. The users were then asked to estimate whether they will like the movie or not and provide an overall rating for each movie by relying solely on the information given in the explanation (lines 2-7). The selection order was randomized to minimize the effects of seeing recommendations or interfaces in a special order. During the user's interaction with the evaluation system, we additionally measured the time needed by the users to submit a rating as to asses the *efficiency* of the user interface. These steps were repeated for all movie/interface combinations.

Some explanation interfaces – such as the tag cloud visualizations – provide information about the content of a recommended movie. This can be problematic when users are able to guess which movie they were rating, which could again lead to an undesired rating bias [Bilgic and Mooney, 2005; Tintarev and Masthoff, 2012]. For this reason, if the users recognized a movie based on the information presented in an explanation, they were instructed to enter this information in the evaluation system and skip to the next recommendation-explanation pair. The ratings for these movie/interface combinations were correspondingly not taken into account in these cases.

After the participants of the study had rated the items based only on the explanations (but without knowing the movie itself), we again presented the recommended movies to the user. This time, however, we disclosed the movie title and as well as detailed information about the recommended movies such as a trailer, the cast overview, the storyline, the plot keywords, and the corresponding genres. The required content data was retrieved from the IMDb Web site[5].

We provided all this detailed information about the movies in order to reduce the time needed for completing the experiment because watching all recommended movies would be too time-consuming. The participants were instructed to read the detailed information about the recommended movies and were then asked to rate the movies again (lines 8-11). According to [Bilgic and Mooney, 2005], from the point of view of an end-user, a good explanation system can minimize the difference between ratings provided in the lines 6 (explanation-based rating) and 10 (rating based on movie information). This way, we were able to measure *effectiveness* by calculating the rating differences. If both ratings coincide, we have a perfect match with a difference of 0. In other words, the explanation helps the user to accurately estimate the quality of an item. Otherwise, if the difference is positive/negative, users will overestimate/underestimate the quality of an item [Bilgic and Mooney, 2005]. Thus, effectiveness is strongly correlated with *persuasiveness*, the ability to persuade a user to buy or try the recommended item. High effectiveness comes together with low persuasiveness and vice versa, leading to a trade-off between these two dimensions [Bilgic and Mooney, 2005].

Note that when collecting explanation and actual ratings in two independent loops, we ensured that participants were able to distinguish between the explanation and the detailed information which caused problems in [Tintarev and Masthoff, 2012].

At the end of the study (line 12), the participants were asked to give feedback on the different explanation interfaces (as to measure user *satisfaction*) by rating the explanation as a whole. Throughout the experiment, we used 1 (lowest) to 7 (highest) rating scales as proposed by [Herlocker et al., 2000] and [Tintarev and Masthoff, 2012]. We additionally measured the user-perceived level of *transparency* for each interface on a 1 (not transparent at all) to 7 (highest transparency) rating scale. The order of the interfaces presented to each user was again randomized to account for learning effects.

Overall, every participant of our study evaluated all 10 explanation interfaces. Each explanation type was used to explain 3 different movie recommendations. Therefore, each participant provided a total of 30 explanation-based ratings and 3 ratings based on detailed information about the recommended movies. The participants could also repeat the experiment. However, almost all of the participants conducted the experiment only once. For each type of explanation interface the explanation ratings of each user were averaged in order to be able to perform the statistical tests used in this study properly.

---

[5]http://www.imdb.com

6.3.2   The underlying tagged movie data set

The evaluation system used in our study is based on a subset of the "MovieLens 10M Ratings, 100k Tags" (ML) data set[6], which consists of movie ratings on a 5-star scale with half-star increments. In addition, the data set contains information about tags that MovieLens users have attached to the individual movies. To the best of our knowledge, this data set is the only publicly available one which contains both rating and tagging data. It contains 10,000,054 ratings and 95,580 tags applied to 10,681 movies by 71,567 users of the online movie recommender service MovieLens. However, no explicit tag preferences are available.

The limited quality of user-contributed tags is one of the major issues when developing and evaluating tag-based recommendation approaches. In [Sen et al., 2007], it was observed that only 21% of the tags in the MovieLens system had adequate quality to be displayed to the user. We therefore selected a subset of the original 10M MovieLens data set because some of the explanation interfaces can only present appropriate explanations if enough data with sufficient quality is available. Consider, for example, a tag cloud containing only two elements. The expressive power and value of such a visualization would be very limited. In addition, we have to ensure that our tag preference estimation algorithm presented in Section 6.2.2 can compute sufficiently accurate predictions. This, however, requires a certain overlap in the tags and the ratings of the movie.

In order to improve the quality of the data set, we therefore applied the following filter operations and constraints on the 10M MovieLens data set in order to delete tags, users, or items for which not enough data was available.

- We removed stop-words from the tags such as "a", "by", "about", and "so", by applying stemming [Porter, 1997] and by filtering out noisy tags, which contain a certain number of characters that are not letters, numbers or spaces, e.g., elements such as smileys[7].

- We required that a tag has been applied by at least 2 different users. This approach was also followed in previous work. In [Vig et al., 2009], for example, Vig et al. require that "a tag has been applied by at least 5 different users and to at least 2 different items".

- We further pre-processed the data by removing items with less than 100 ratings and less than 10 tags, of which at least 7 must be different.

These steps also ensure a sufficient recommendation quality of the nearest-neighbor collaborative filtering algorithm and – as a side effect – reduce its computation time, which is important in our experiment since we have to compute a user's nearest neighbors and his or her personalized recommendation list online. Table 6.2 shows the MovieLens 10M data set characteristics before and after the application of the filter operations and constraints.

Data Set	#Ratings	#Users	#Items	#Tags
MovieLens 10M	10,000,054	71,567	10,681	95,580
MovieLens 10M Subset	5,597,287	69,876	963	44,864

Table 6.2: Data set characteristics. We used a subset of the MovieLens 10M data set in our experiment.

A limitation of this study is that we only focus on popular items containing overall more than 50% of the available ratings and 45% of the available tag assignments. This leads to a new bias towards the popular items in the data. However, as described above in Section 6.3.1, focusing on popular movies is advantageous for the experimental design.

6.3.3   Hypothesis development

We tested four hypotheses. In our previous study presented in the last chapter, we stated that due to the visual nature of a tag cloud, users can grasp the content information inside a cloud very fast. Furthermore, we showed that the decision time when using the tag cloud explanations is significantly

---

[6]http://www.grouplens.org/node/73

[7]A smiley is a sequence of characters, such as punctuation marks and letters, representing emoticons.

shorter than for the keyword-style explanation [Bilgic and Mooney, 2005] which has a more complex tabular structure. We believe that most of the users are already familiar with the idea of tag clouds and know the basic rationale behind a tag cloud and thus will perceive them as easier to understand. Unlike Herlocker et al.'s explanation methods presented in Section 6.2.1, they also convey more data about an item's content. Overall, our assumption is therefore that by using tag cloud explanations users can decide at least equally fast as with Herlocker et al.'s explanation methods. For these reasons we hypothesized that:

Hypothesis H1. *The tag cloud explanations enable users to make decisions as fast as the other explanation interfaces tested in the study.*

The results of our previous study also revealed that users can make better decisions when using the tag cloud interfaces rather than with keyword-style explanations. As users are already familiar with the concept of tag clouds, we assume that tag cloud explanations are easier to interpret. Unlike Herlocker et al.'s explanation methods, they also convey content data about the recommended item (see Table 6.1). Therefore, users can make a more precise estimate of the quality of the recommended item. Furthermore, we believe that personalization is a helpful means to further improve a user's evaluation of the recommended item which was also the intuition of [Tintarev and Masthoff, 2012] (compare with Tintarev and Masthoff's first hypothesis, H1). Thus, we hypothesized that:

Hypothesis H2. *Personalized tag cloud explanations are more effective than non-personalized tag clouds and the other explanation interfaces tested in the study.*

As described in Section 6.2.2, we further improved the interface perstagcloud to make the rationale more transparent for the user. Unlike the other explanations, we believe that by using tag preferences, perstagcloud can help to establish a connection between the user's past rating behavior and the recommended item, leading to a more transparent user experience. We therefore hypothesized that:

Hypothesis H3. *Users perceive the personalized tag cloud explanations as more transparent than non-personalized tag cloud and the other explanation interfaces tested in the study.*

Since we assume that users can make better decisions faster when using the tag cloud interfaces, we believe that users will prefer them over the others. In particular, our assumption is that users will reward the perstagcloud interface because it is personalized to the users' interests. Tintarev and Masthoff also hypothesized that bringing personalization into their feature-based explanations will make the users more satisfied [Tintarev and Masthoff, 2012] (compare with their second hypothesis, H2). Therefore, our final hypothesis was:

Hypothesis H4. *Users are more satisfied with personalized tag cloud explanations compared to non-personalized tag clouds and the other explanation interfaces tested in the study.*

6.3.4 Participants of the study

We recruited 105 participants from ten different countries for our study. Instead of focusing on students at our institution alone, we tried to recruit users from different demographic groups to achieve more representative results. Still, most users were generally interested in movies and had at least high-school education. The average age was 28 years. More details about some of the demographic characteristics of the participants are given in Table 6.3.

6.3.5 Data collected in the experiment

During the experiment, the participants provided a total of 2,370 overall movie ratings. On average, each user provided 23 movie ratings. Because participants were also allowed to repeat the experiment, we collected 3,108 ratings for the explanations (on average 30 ratings per user). Figure 6.4 shows the distribution of the movie ratings collected in our study. It can be seen that users preferred to rate movies they liked, i.e., a *positivity bias* can again be observed which is in line with the findings of previous work [Marlin et al., 2007; Vig et al., 2010].

Gender	Female	52 (49.52%)
	Male	53 (50.48%)
Education	A-level	58 (55.24%)
	Bachelor	5 (04.76%)
	Master	32 (33.33%)
	PhD	2 (01.90%)
	Other	5 (04.76%)
Nationality	10 countries	(Germany, Turkey, China, etc.)
Age	19-23	34 (32.38%)
	24-28	29 (27.62%)
	29-33	27 (25.71%)
	34-38	11 (10.48%)
	39-45	4 (03.81%)
Interest in movies	Low	14 (13.33%)
	Normal	48 (45.71%)
	High	43 (40.95%)

Table 6.3: Demographic characteristics of participants (total 105).

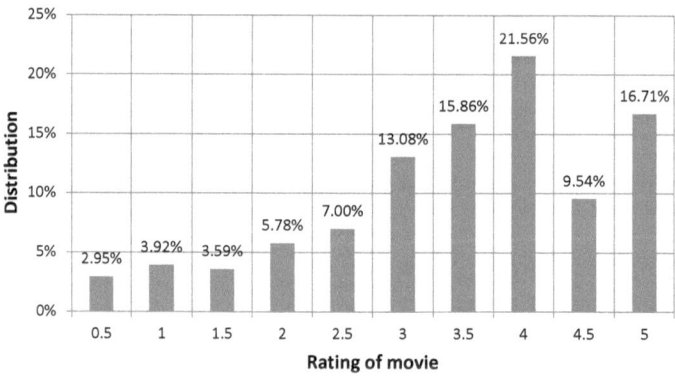

Figure 6.4: Distribution of movie ratings on a 5-point rating scale with half-point increments.

6.4 Results and discussion

SPSS 20 was used for data analysis and all the tests were done at a 95% confidence level. Regarding the statistical methods applied in our analysis, we used the Friedman test throughout this work to analyze whether observed differences between the various interfaces are statistically significant. We also used t-tests and repeated-measures ANOVA, but we will only report the results of the non-parametric Friedman test here since the results were similar across all tests. The Friedman test is suggested by [Demšar, 2006] for a comparison of more than two systems. Once a significant difference was found, we applied the post-hoc Wilcoxon Signed-Rank test to identify where the differences are. In order to interpret our Wilcoxon test result, a Bonferroni correction was accordingly applied and thus all the effects are reported at a $p < 0.005$ level of significance, if not stated otherwise. Detailed test statistics are listed in Appendix B.2.

6.4.1 Efficiency

Efficiency stands for the ability of an explanation interface to help the users make their decisions faster [Tintarev and Masthoff, 2011]. We defined the decision time in our scenario as the time needed by the users to submit a rating after seeing the corresponding explanation[8].

Figure 6.5 shows the mean time (in seconds) needed by the users for each interface type. The standard deviation values are depicted in Appendix B.2.1. We can see that users interacting with the tag cloud interfaces need the most time for decision making, whereas the most efficient interface (rated4+) needed less than half of the time. Thus, hypothesis H1 is rejected. We assume that users interact longer with explanation interfaces which provide more information such as content data. Since the tag cloud interfaces are the only ones which exploit content data and which need significantly more time than the other explanation types, we believe that the difference is due to the provision of content data. The tag cloud interfaces contain several keywords which characterize a movie, whereas the interface rated4+ simply shows the percentage of ratings for an item which are equal or higher than 4 in one single string. Users interacting with **perstagcloud** need the most time as it is the only interface which provides both personalization *and* content data (see Table 6.1). However, no significant difference between the two tag cloud interfaces could be found.

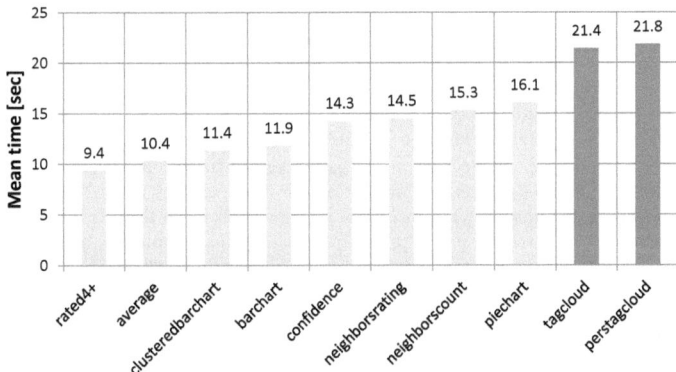

Figure 6.5: Mean time for submitting a rating after seeing the corresponding explanation. Light gray bars indicate explanations with a mean time significantly different from the base case **perstagcloud** ($p < 0.005$, $N = 291$).

While the data clearly shows that some interfaces lead to faster decisions, the interesting question addressed later on in this work is whether and to which extent the efficiency of an explanation interface can actually help to improve user satisfaction.

6.4.2 Effectiveness and persuasiveness

Effectiveness is the ability of an explanation to help a user to assess accurately the (high or low) quality of the recommended item [Tintarev and Masthoff, 2011]. Effective explanations help users of a recommender system make better decisions [Bilgic and Mooney, 2005].

We approximate the effectiveness of an interface by using Bilgic and Mooney's effectiveness metric [Bilgic and Mooney, 2005] (see Section 6.3.1). Figure 6.6 shows the mean difference values of the interfaces.

---

[8]The experiment was conducted over the Internet. Therefore, timings may be unreliable and may impact efficiency results.

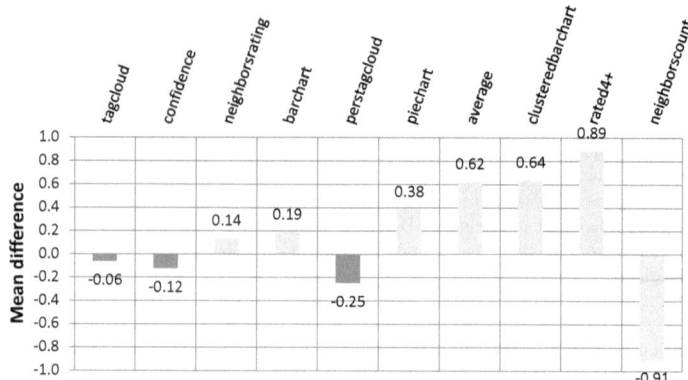

Figure 6.6: Mean difference of explanation-based ratings and ratings given based on more detailed information. Light gray bars indicate explanations with a mean difference significantly different from the base case `perstagcloud` ($p < 0.005$, $N = 291$).

The results indicate that the `tagcloud` interface leads to a very small difference between the explanation-based rating and the rating based on detailed information, which can be interpreted as a sign of good effectiveness and low persuasiveness. The `neighborscount` interface, on the other hand, leads to the effect that users underestimate how they would like a movie when seeing only how many like-minded users (neighbors) rated the movie. Thus, the probably undesired effect of persuasiveness would be high.

It is important to note that observing a mean value close to 0 alone is not sufficient for identifying effective explanation interfaces correctly. Consider a case where an explanation interface always leads to the effect that users give ratings which are close to the overall mean rating value. While the actual ratings follow some given distribution and probably deviate strongly from the mean rating, the mean difference between explanation ratings and actual ratings can also be 0. Therefore, we have to take such *averaging effects* into account [Bilgic and Mooney, 2005; Tintarev and Masthoff, 2008a].

In [Tintarev and Masthoff, 2012], the issue of the "neutral middle of the scale ratings" is also discussed. The authors do not force the participants to make a decision. Instead, they allow the users to skip a rating and analyze the opt-out ("no opinion") rates later on.

In this study we follow the approach of [Bilgic and Mooney, 2005] which reduces the risk of averaging effects. Therefore, beside the mean difference values shown in Figure 6.6, we also have to consider the correlation of the explanation-based ratings and the actual ratings as well as the corresponding standard deviation values. Thus, according to [Bilgic and Mooney, 2005], from the user's point of view, a good explanation interface has (a) a mean difference value of 0, (b) a high positive correlation between explanation ratings and actual ratings, and (c) a low standard deviation value.

We show correlation and standard deviation values in Table 6.4. The triangles in the subsequent tables indicate how the tables are sorted. In Table 6.4 the data rows are sorted in descending order of the correlation value.

Figure 6.6 shows that the mean differences of the tag cloud interfaces are close to 0, which – as mentioned above – is an indication that the interfaces help users accurately estimate the quality of an item. Moreover, from Table 6.4, we can see that the tag cloud interfaces in addition have the highest correlation values and the lowest standard deviation values. The explanation interfaces `confidence`, `neighborsrating`, and `barchart` also have mean difference values close to 0. However, `tagcloud` has a correlation value which is twice as high as well as a lower standard deviation value.

Contrary to our intuition, the results indicate that `tagcloud` explanations are more effective than their personalized counterpart (`perstagcloud`), that is, there is no support for hypothesis H2. Tintarev and Masthoff also observed that contrary to expectation, personalization was detrimental to effectiveness

#		Pearson Corr ▽	Std Dev
1	`tagcloud`	0.506	1.570
2	`perstagcloud`	0.504	1.661
3	`confidence`	0.243	1.904
4	`piechart`	0.239	2.002
5	`neighborsrating`	0.231	1.867
6	`average`	0.226	1.772
7	`barchart`	0.193	1.903
8	`clusteredbarchart`	0.192	2.071
9	`rated4+`	-0.049	1.960
10	`neighborscount`	-0.053	2.182

Table 6.4: Pearson correlation values between explanation ratings and actual ratings and standard deviation values of the mean differences.

[Tintarev and Masthoff, 2012]. In their work, which was developed in parallel to our own work, the authors observed in three experiments in two domains that their method of personalization hindered effectiveness, but increased satisfaction with explanations; this also holds for the tag cloud interfaces evaluated in our study. We will discuss the results for satisfaction later on in Section 6.4.4. We believe that one reason for the difference regarding effectiveness might be the fact that we relied on estimated tag preferences in this study. In future work, alternative heuristics to estimate the tag preferences can be used as to further increase the effectiveness of the personalized interface.

Regarding the aspect of persuasiveness, explanation interfaces which lead to mean values above 0 in Figure 6.6 cause the user to overestimate the actual rating (and real value of an item). Such interfaces could be used in situations where an item should be promoted. The `rated4+` interface, for example, causes the user to overestimate the actual rating by 0.89 on average, which we consider to be comparably high given our 7-point rating scale. Analogously, the explanations below 0 cause the user to underestimate the quality or value of an item.

In some domains such as finance and tourism, using interfaces that lead to an overestimate of an item's quality could be risky. In the long term, a recommender system that tries to continuously persuade the customer toward certain items may leave the user sooner or later with the impression that the system is cheating because the system overstates the advantages or value of the items. Thus, we assume it to be a comparably "safe" strategy, when we are able to keep the persuasiveness level within a certain range or slightly on the negative side. Tintarev and Masthoff, for instance, found that users perceived overestimation to be less helpful than underestimation [Tintarev and Masthoff, 2008b]. Overall, we argue that it is important to choose the direction of the persuasiveness depending on the current recommendation scenario and goals as well as depending on the general business strategy.

Our study shows that the newly proposed tag cloud interfaces are among the most effective explanation interfaces which at the same time have a very small tendency to cause the user to over- or underestimate the real value of a recommended item.

The tag cloud interfaces are the only interfaces in our study which make use of domain specific content information. We consider this as being an indication that users are able to evaluate explanations based on content information more precisely. Note that among the best performing methods in the study by [Herlocker et al., 2000] was the "Favorite actor or actress" interface, which also presents content information to the user. By providing content data users are able to take advantage of their personal experience and knowledge in the considered domain. Thus, they are able to estimate the quality, value, or relevance of an item more precisely. Overall, our design suggestion is therefore:

Guideline 1. *Use domain specific content data to boost effectiveness.*

Since previous studies [Herlocker et al., 2000; Bilgic and Mooney, 2005] did not differentiate between explanations with and without content data, content-based explanations were not specifically chosen in the selection process in Section 6.2 either. We see the systematic analysis of the effects of different amounts and types of content data in explanations as one of the next steps in future work.

6.4.3 Transparency

From the user's point of view, a recommender without an explanation facility is often a black box which accepts a user profile as input and returns a personalized recommendation list as output [Herlocker et al., 2000]. One of the main goals of explanations is to reveal the recommendation logic within the black box in order to make the system more transparent [Tintarev and Masthoff, 2011]. A transparent explanation interface helps the user to understand why certain items were recommended and potentially gives the user a general idea of how the system works.

We measured the user-perceived level of transparency by asking the user whether the considered interface helps to understand how the recommender system works. The users had to rate each interface on a scale from 1 (not transparent at all) to 7 (highest transparency). The results are depicted in Table 6.5.

#		transparency ▽	Std Dev
1	perstagcloud	5.61	1.53
2	barchart	5.51	1.26
3	piechart	5.41	1.21
4	clusteredbarchart	5.40	1.25
5	rated4+	5.27	1.28
6	neighborsrating	5.12	1.13
7	average	5.07	1.46
8	tagcloud	**5.05**	1.60
9	confidence	**4.65**	1.34
10	neighborscount	**2.80**	1.70

Table 6.5: Mean response of the users to each explanation interface regarding transparency, based on a rating scale from 1 to 7. Numbers printed in bold face indicate the explanation interfaces with a mean response which is significantly different from the base case perstagcloud ($p < 0.005$, $N = 105$).

The results support our third hypothesis, H3. The novel perstagcloud interface was perceived by the users to be the most transparent one. We believe that the additional window which appears when the user moves the mouse over a tag is particularly helpful for the user to understand the recommendations. Notice the comparably high standard deviation of the perstagcloud interface. This effect may be a consequence of the fact that some users did not agree with the automatically estimated tag preference values. We therefore believe that a higher transparency value along with a lower standard deviation value can be achieved by improving the tag preference inference algorithm described in Section 6.2.2.

Interestingly, the non-personalized tagcloud interface is perceived to be significantly less transparent than the perstagcloud interface. A closer look at the individual behavior of users showed that users who have rated both types of tag cloud interfaces at the end of the experiment had a tendency to evaluate them in a side-by-side comparison. Several times the users exhibited an affinity toward one interface by significantly reducing the transparency value for the other one when they had seen both of them. Although the order of the interfaces presented to each user was randomized, users could re-rate an interface at any time. Thus, many users "corrected" the transparency value for tagcloud after rating its personalized counterpart.

Although there was no significant difference between average and perstagcloud, the trend was that the average interface was perceived less transparent. This trend was confirmed by participant comments who stated that a threshold rating value for this interface was missing. Therefore, users did not know which rating value was used for classifying items as being worth recommending or not.

The simple string based interface neighborscount performed by far worst. Users obviously did not understand how the number of neighbors who provided a rating for the target item was used in the recommendation process.

6.4.4 Satisfaction

Explanation interfaces are an important medium of communication between recommender system and user and can convey valuable information to the user. Thus, a good explanation interface can increase a user's overall satisfaction with a system or his or her acceptance of a system [Herlocker et al., 2000; Cosley et al., 2003; Tintarev and Masthoff, 2012]. In our questionnaire at the end of the experiment we explicitly asked users about their overall satisfaction with the explanations. Table 6.6 shows the mean response on overall satisfaction of 105 users to each explanation interface based on a scale of 1 to 7.

#		satisfaction ▽	Std Dev
1	perstagcloud	4.96	1.93
2	average	4.70	1.39
3	rated4+	4.63	1.50
4	tagcloud	4.59	1.91
5	clusteredbarchart	4.57	1.60
6	barchart	4.56	1.40
7	confidence	4.45	1.39
8	piechart	4.32	1.75
9	neighborsrating	**3.95**	1.46
10	neighborscount	**2.09**	1.38

Table 6.6: Mean response of the users to each explanation interface regarding satisfaction, based on a rating scale from 1 to 7. Numbers printed in bold face indicate the explanation interfaces with a mean response which is significantly different from the base case `perstagcloud` ($p < 0.005$, $N = 105$).

The highest mean rating value of `perstagcloud` indicates that the participants of our study preferred the `perstagcloud` approach over other types of explanations which also supports our last hypothesis, H4. The trend is that the personalization of the tag cloud interface improves user satisfaction, while personalization is detrimental to effectiveness as described above. The users made better decisions with the help of `tagcloud` explanations, but preferred the personalized explanations of `perstagcloud`. Our findings are in line with the findings of the very recent work of Tintarev and Masthoff [Tintarev and Masthoff, 2012].

Among the other well-performing explanation types with respect to user satisfaction were the `average`, `rated4+`, and `tagcloud` interfaces. The `average` interface was rated highly by the users, while it performed worst in the study of [Herlocker et al., 2000] more than ten years ago. However, remember that [Herlocker et al., 2000] evaluated the interfaces from a promotion perspective (persuasiveness) only. Our study provides additional evidence that it is important to evaluate explanation interfaces from different points of view [Tintarev and Masthoff, 2012]. When selecting an explanation interface, one has therefore to be aware of the different types of effects of explanations and align them properly with the business strategy.

An analysis of the successful approaches with respect to user satisfaction lets us develop another design guideline. Explanation concepts, which the users already know, e.g., from typical Web 2.0 interfaces, are preferred over others. Tag clouds, for example, are nowadays a frequently used visualization and interaction technique on many popular Social Web sites such as Delicious and Flickr[9]; similarly, the `average` interface is commonly used on various Web sites such as IMDb or Amazon.com. One participant of our study, for example, commented that *"I usually base my decision whether or not to rent a movie on the movie's overall average rating information presented on the IMDb Web site."*. However, some of the participants of our experiment had not seen tag clouds before, which we see as the main reason for their comparably high standard deviations. Based on these findings, we suggest:

Guideline 2. *Use explanation concepts the user is already familiar with, as they require less cognitive effort and are preferred by the users.*

Up to this point, we have analyzed the different aims of explanations separately. Before we discuss the aspect of *trust*, which we do not measure directly, we will report the results of an analysis of the

---

[9]http://www.delicious.com, http://www.flickr.com

relationships between the different factors. In particular, we will analyze the impact of efficiency, effectiveness, and transparency on user satisfaction since satisfaction is often used as a dependent variable in recommender system research [Cremonesi et al., 2011].

6.4.5   Relationships between variables

The user-centric evaluation of the causal relationships between different evaluation factors is a recent topic within recommender system research [Pu et al., 2012]. We conducted a *path analysis* to analyze the relationships between the different quality factors for recommender system explanations. SPSS AMOS 20 was used for model building and path analysis. Path analysis was done using regression analysis, which is a standard approach in social sciences and in educational research [Tuijnman and Keeves, 1994] to study of potential causal relationships.

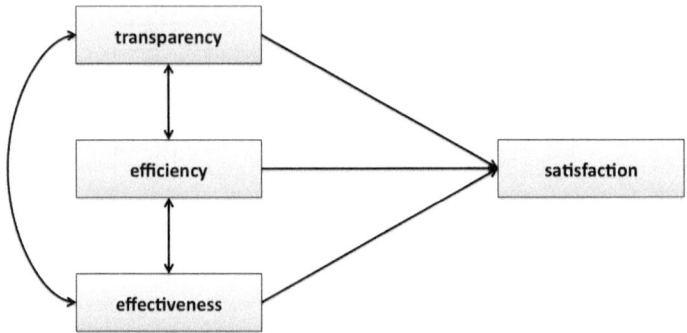

Figure 6.7: The SPSS AMOS path analysis model which describes the dependencies among the variables.

Figure 6.7 shows our model which we used as input for the SPSS AMOS path analysis tool. Each quality factor is represented by an independent or dependent variable in the model. We see satisfaction as the dependent variable which depends on the quality factors efficiency, effectiveness, and transparency, which are consequently modeled as independent variables. The edges between the independent variables represent the covariance parameters to estimate. We conducted a path analysis for each explanation interface separately. The results of each path analysis run are shown in Table 6.7 and Table 6.8.

	transparency	efficiency	effectiveness	R^2
		$\downarrow$		
		satisfaction		
clusteredbarchart	**0.760**	-0.003	0.108	0.350
barchart	**0.773**	-0.005	0.144	0.467
neighborsrating	**0.547**	0.001	0.112	0.187
confidence	**0.509**	-0.002	0.066	0.254
neighborscount	**0.473**	0.008	0.171	0.467
rated4+	**0.674**	0.016	**0.302**	0.469
average	**0.482**	0.010	0.235	0.339
tagcloud	**0.888**	0.003	0.237	0.510
perstagcloud	**0.883**	0.002	0.238	0.505
piechart	**0.705**	-0.011	0.317	0.330
∅	0.669	0.002	0.193	0.388

Table 6.7: Maximum likelihood estimates of the regression weights. Bold figures indicate weights with a significant effect ($p < 0.001$, $N = 291$).

Table 6.7 shows the maximum likelihood estimates of the regression weights, along with their accompanying R squared values, indicating the goodness of fit for estimating the parameters of these models. The results show clearly that transparency – independent of the used interface – has a significant positive effect on user satisfaction. Because both transparency and satisfaction use a 7-point Likert scale (from "not at all" to "a lot"), we have also considered the Pearson correlation coefficient and the Spearman's rank correlation coefficient between transparency and satisfaction which are 0.84 and 0.57 respectively. We see this as a strong indication that users are generally more satisfied with explanation facilities which provide justifications for the generated recommendations. This even holds when the user's beliefs of how the system works are actually wrong. Our design suggestion is therefore:

Guideline 3. *Increase user-perceived transparency through explanations in order to increase user satisfaction.*

When we look at the path analysis results for efficiency, we can observe that decision time – in contrast to transparency – seems to have no influence on user satisfaction. The average regression weight of efficiency is close to 0. Based on this observation, we can derive a further design suggestion for explanations:

Guideline 4. *Explanations should not primarily optimized for efficiency. Users take their time for making good decisions and are willing to spend the time on analyzing the explanations.*

Furthermore, we assume that trying to decrease decision time by providing less information can even lead to a negative effect on user satisfaction. This hypothesis has however to be validated in an additional study as our data does not contain sufficient evidence so far.

With respect to the relation of effectiveness and user satisfaction, we see that the average weight for effectiveness is less than one-third of the average weight for transparency. Except for the `rated4+` interface, effectiveness had no significant (short term) effect on user satisfaction in our study. However, we believe that effectiveness has a long-term effect which we were not able to capture in our single-session experiment and which would require long-term analysis spanning longer periods of time and multiple sessions. In our view, effectiveness is a crucial factor for the long-term success of an explanation interface and recommender system as a whole. Consider, for example, the `average` interface. The interface performs comparably poor with respect to effectiveness, see Figure 6.6, but led to good user satisfaction as shown in Table 6.6. We asked users who usually base their decision on IMDb's overall average rating interface to which extent the interface meets their expectations. One user commented that *"Often I am very surprised about the low average rating value of a movie."*, which we see as an indicator that in the long run, the user probably will loose his trust in this simple quality indicator provided by the system. Due to the limitations of our experiment, we only have limited supporting evidence from the data so far. Based on these considerations, our last design suggestion is:

Guideline 5. *Enhance effectiveness in order to increase user satisfaction in the long run.*

	transparency $\updownarrow$ effectiveness		transparency $\updownarrow$ efficiency		efficiency $\updownarrow$ effectiveness	
	Cov	Corr	Cov	Corr	Cov	Corr
clusteredbarchart	-0.187	-0.123	-0.415	-0.054	-0.508	-0.067
barchart	-0.338	-0.217	-0.562	-0.072	0.241	0.031
neighborsrating	-0.014	-0.010	-1.520	-0.560	-0.622	-0.021
confidence	0.249	0.145	-0.590	-0.016	0.950	0.026
neighborscount	0.820	0.292	0.924	0.044	4.877	0.238
rated4+	0.356	0.208	-0.473	-0.059	0.107	0.013
average	0.326	0.195	0.626	0.058	0.194	0.023
tagcloud	-0.052	-0.031	-1.743	-0.090	0.601	0.044
perstagcloud	-0.054	-0.032	-1.741	-0.090	0.601	0.044
piechart	0.162	0.101	-0.825	-0.082	-0.622	-0.055
$\emptyset$	0.127	0.053	-0.632	-0.092	0.582	0.028

Table 6.8: Covariances and correlations among the independent variables.

Table 6.8 finally contains the covariance and correlation values between the independent variables.

As expected, the average correlation values of the independent variables are close to 0, i.e., there is no linear correlation between the variables. However, for some interfaces, we can observe trade-offs between different dimensions. For example, a low negative correlation value of -0.56 exists between transparency and efficiency for the interface **neighborsrating**, an interface that displays a tabular view of the ratings within the user's neighborhood. Tintarev and Masthoff discuss this trade-off and state that high transparency may impede efficiency since the users would spend more time with an explanation [Tintarev and Masthoff, 2007b].

Overall, the results confirm our decision to model efficiency, effectiveness, and transparency as independent variables since we could not detect interdependencies among them in general.

6.4.6 Implications for trust

As mentioned in Section 5.2, system trust focuses on the long-term relationship between a user and a recommender [O'Donovan and Smyth, 2005b]. In the context of recommender systems, trust can be seen as the willingness of users to believe in the appropriateness of the recommendations [Cramer et al., 2008], i.e., trust is a measure for how credible and reliable a system is.

Already [Tintarev and Masthoff, 2007a] mention the difficulty of measuring trust and propose measuring it through user loyalty and increased sales. Although there exist user studies which try to make conclusions on the formation of user trust [Pu and Chen, 2006, 2007], we believe that large scale online experiments with real user populations who interact with the system are appropriate for this task. In such online evaluations the behavior of users can be measured over time. As such online evaluations are expensive and risky, we will try to analyze the effects of explanations on trust in a theoretical way based on the different findings reported in this work so far and by summarizing results of previous research in that context.

We consider satisfaction as a prerequisite for trust. Satisfaction with the explanations increases the overall satisfaction with the system [Cosley et al., 2003]. A user who is not satisfied with the system is not likely to develop trust in a system. Trust is necessary to keep the user satisfaction sustained over a long period of time.

We showed that transparency is a highly important factor for user satisfaction. Thus, we believe that transparency is also an important factor for trust, which is also suggested in the literature. Swearingen and Sinha, for example, mention that transparency is an important factor which strongly affects the user's trust in a system [Swearingen and Sinha, 2002].

Efficiency, on the other hand, does not appear to be particularly important for user satisfaction. In our study we did not analyze the long-term effects of efficiency on satisfaction. However, we believe that efficiency has a limited effect on trust. An efficient system might be more comfortable to use since it requires less cognitive effort but such a system may not necessarily be more trustworthy.

Bilgic and Mooney argue that effectiveness is more important than persuasiveness in the long run as greater effectiveness can help to establish trust and attract users [Bilgic and Mooney, 2005]. In our experiment we could not measure a significant effect of effectiveness on user satisfaction (except for the **rated4+** interface). However, we believe that in a long-term user study the importance of this factor will be visible. Explanations that continuously lead to an overestimate of an item's quality can be risky since after a while the user may get the impression that the recommender system is cheating because from the user's point of view the recommender is promoting items without taking the user's true preferences into account. On the other hand, if the explanations continuously lead to an underestimate of an item's quality, the system may leave the user eventually with the impression that the system is not intelligent enough to generate appropriate recommendations. Thus, both positive and negative persuasiveness can cause the loss of trust to users.

Putting the thoughts together, we conclude that in particular the newly proposed tag cloud interfaces represent examples for explanations which can help increase the user's trust in a system. They are effective, transparent, and improve the users' satisfaction with the explanation.

6.5 Summary

In this chapter, we have presented the results of a user study in which ten different explanation types were evaluated with respect to several evaluation factors. Beside explanation interfaces from the literature, we included our newly proposed visualizations based on tag clouds and presented an algorithm for estimating the user's preference toward certain aspects of a recommendable item, i.e., a user's tag preference values. Our study provides additional evidence that it is important to consider the different evaluation aspects of explanation interfaces together when selecting the right explanation interface for a recommender.

Since explanations represent the interface between a user and a recommender, they play a crucial role for the success of the recommender system. The results additionally show that our newly proposed personalized tag cloud interface is perceived as highly transparent by the users. In addition, the results show that tag cloud explanations increase user satisfaction even though they demand higher cognitive effort from the user. We believe that these characteristics are necessary to establish user trust in a recommender system. Besides an analysis of different effects, a first analysis of the interdependencies between the different aims of explanations was provided. Based on these insights and other observations of the study, we introduced a set of possible design suggestions regarding explanation interfaces for recommender systems.

The evaluation of more variations of the tag cloud interface design followed by a discussion of ways to further improve the user interface and a comparison with very recently proposed interfaces represent possible directions for future work. The limitations of this study will be discussed in the next chapter.

Overall, we see this study as a further step toward a better understanding of how to build user interfaces for recommender systems that support the user in the decision making and buying process in the best possible way.

Chapter 7

Summary and perspectives

The goal of recommender systems is to provide personalized recommendations of products or services to users facing the problem of information overload on the Web [Adomavicius and Tuzhilin, 2005]. They provide personalized recommendations that best suit a customer's taste, preferences, and individual needs. Especially on large-scale Web sites where millions of items such as books or movies are offered to the users, recommender system technologies play an increasingly important role. One of their main advantages is that they reduce a user's decision-making effort [Felfernig et al., 2011; Ricci et al., 2011b]. However, recommender systems are also of high importance from the service provider or system perspective. For instance, they can convince a customer to buy something or develop trust in the system as a whole which ensures customer loyalty and repeat sales gains [Jannach et al., 2010; Ricci et al., 2011a].

In traditional collaborative filtering systems, the assumption is that customers provide *one* overall rating for the items which they have purchased. Nowadays customers in some domains are given the opportunity to provide more fine-grained feedback. In the tourism domain, for example, customers can rate hotels along quite a number of different dimensions such as cleanliness, breakfast service, value for money, or staff friendliness [Jannach et al., 2012]. However, static rating systems with predefined dimensions are outdated because professional experts are needed for design; in addition, new dimensions may emerge over time that were not covered by the predefined and pre-thought dimensions designed or foreseen by a domain expert [Shirky, 2005]. We argue that users want to use their own rating system making the development of predefined rating systems by domain experts obsolete. For example, think about users who want to state in their own words, how funny a movie is, without the limitations of a static rating system.

Therefore, we introduced the concept of *rating items by rating tags* in this work. The idea was to apply Shirky's thoughts about ontological classification and static categorization schemes [Shirky, 2005] to rating dimensions. We extended the usual user-item rating matrix not only by a set of user-provided tags for the items, but also by *tag preferences* describing the user's opinion about the item features represented by these tags. For example, a user assigns the tag *funny* to a movie and rates the tag in order to express *how* funny the movie is. Thus, users can express themselves better through tag preferences. Additionally we are able to build a more detailed profile about the user and his or her interests. Of course, the free-form labeling and rating of items by everybody comes at the price of a less homogeneous and more unstructured set of item annotations and rating dimensions. However, as stated in the introduction, the amount of user contributed information provided by the Social Web poses both new possibilities and challenges for recommender systems, which were addressed in this thesis.

The following section briefly summarizes our contributions and achievements in the field of leveraging tagging data for recommender systems. We will also discuss the limitations of the results presented in this thesis which lead to suggested directions for future work.

7.1 Contributions and limitations

We can divide the contributions and achievements of this work into two main parts: tag-based recommendations (Chapters 3 and 4) and tag-based explanations (Chapters 5 and 6).

In Chapter 3, we presented a new tag recommender algorithm called LocalRank. LocalRank is a graph-and-neighborhood-based tag recommendation approach which is inspired by the popular FolkRank algorithm [Hotho et al., 2006]. A major problem of FolkRank is that it does not scale to larger problem sizes, which is crucial for real-world scenarios. Rank computation and weight propagation in LocalRank is done in a similar way to FolkRank but without iterations. As the name suggests, LocalRank computes the rank weights based only on the local "neighborhood" of a given user and resource. Instead of considering all elements in the folksonomy, LocalRank focuses on the relevant ones only. Thus, LocalRank can significantly reduce the time needed for computing the recommendations while maintaining or slightly improving recommendation quality. The evaluation on the commonly-used Delicious data set showed that LocalRank can generate tag recommendations in real-time also for larger data sets, whereas FolkRank requires more than 20 seconds for generating one single recommendation list using the same hardware configuration. Yet, the accuracy of our method is on a par with or slightly better than FolkRank.

One limitation of the work relating to LocalRank is that only the Delicious data set, which is a commonly-used data set in related work, was used for evaluating the algorithm. It is desirable to validate LocalRank on data sets of other social tagging platforms in order to decide whether it is sufficient to consider only the neighborhood of a given user-resource recommendation query. We are also aware that only FolkRank was used for comparison. Recently, new algorithms have been proposed which outperform FolkRank's predictive accuracy on certain data sets. However, we still see FolkRank as one of the state-of-the-art techniques for tag recommendation and use it as a baseline for comparison because most current literature refers to it as a baseline. The availability of the source code is another reason to choose FolkRank as the baseline approach as it ensures a fair comparison between algorithms.

In Chapter 4, we first introduced our new approach of rating items by rating tags. Users can define their individual rating dimensions by assigning tags to the item to be rated. Note that the user can either create new tags or select tags from a list of recommended tags, which can be provided by our tag recommender algorithm LocalRank presented in Chapter 3. We then showed in Chapter 4 and the following chapters how these tag preference values can be exploited for recommender systems.

In Chapter 4, we further proposed first algorithms that consider tag preferences to generate more accurate predictions. We also provided a way to automatically infer tag preference values from the items' overall ratings because tag preference values are not available yet. The evaluation on two different data sets revealed that the prediction accuracy of our recommendation scheme which exploits tag preference data is better than previous tag-based recommender algorithms and recent tag-agnostic matrix factorization techniques. Note that we conducted experiments with both real and estimated tag preference data. The results showed that even when using estimated tag preference data, our tag-based recommendation scheme outperformed the other algorithms.

We see limitations in the experiments with *real* tag preference data. The data set consisting of real item-specific tag preferences, which were collected in our first user study described in Chapter 5, is rather small, compared to the existing evaluation data sets such as the MovieLens data set. However, since our approach of rating items by rating tags is a new approach, no data set which contains tagging data together with tag preference values was available. Therefore, we had to use the data collected by ourselves. The next step is to validate our methods on larger data sets with real tag preference values. We assume that in future such data will be available on Social Web platforms. Note that our data set with real tag preference values only contains manually selected high-quality tags which is another limitation of this work. In subsequent steps, new tag quality metrics need to be defined which automatically distinguish tags that are appropriate for algorithms that exploit tag preference data, from those which are not, e.g., ambiguous tags.

Limitations on the generality of the results is another issue which is also present in the evaluation of the LocalRank algorithm. Strictly speaking, the evaluation results are only valid on the used data sets. This is a general issue of offline experiments which is known to the recommender system community. However, we believe that the results can be generalized to other data sets and domains, that is, the ranking of the algorithms should be consistent.

In Chapter 5, we analyzed the effects of three explanation interfaces on the quality dimensions efficiency, effectiveness, persuasiveness, satisfaction, and trust. We compared two newly proposed interfaces based on personalized and non-personalized tag clouds with keyword style explanations proposed in previous work. In order to personalize the explanations, the personalized tag cloud interface makes use of tag preference data which was introduced in Chapter 4. Colors are used to indicate whether the user will like, dislike, or feel neutral about the item features represented by the tags in the cloud. The results of our first user study with 19 participants showed that users can make better decisions faster when using the tag cloud interfaces rather than the keyword-style explanations. In addition, users generally favored the tag cloud interfaces over keyword-style explanations.

However, we could only recruit a small number of participants for our first study because the experiment was time-consuming since multiple sessions per user were necessary to finish the experiment. Another limitation is that users were not able to assign new tags to movies because we wanted to ensure that we have a reasonable overlap in the used tags. However, the fact that we have applied appropriate statistical tests compensates for the first limitation and strengthens the reliability of the results. We see the second limitation as a consequence of the first limitation, which was addressed in our follow-up user study.

Therefore, in Chapter 6, we changed the experimental design and conducted a broader user study which involved more participants than our first study described in Chapter 5. We recruited 105 subjects and analyzed the effects of ten different explanation interfaces. Among them were the two tag cloud interfaces from our previous study. We additionally included the evaluation factor user-perceived transparency to the set of evaluation dimensions. From the results, it can be seen that, in particular, our newly proposed tag cloud interfaces can help increase the user's trust in a system. They are effective, transparent, and improve user satisfaction even though they demand higher cognitive effort from the user.

In Chapter 6, we also presented first insights on the interdependencies between the different evaluation factors of explanations. We analyzed the influence of the quality factors efficiency, effectiveness, and transparency on user satisfaction. Based on these new insights, we then derived a first set of possible guidelines for designing and choosing an explanation interface for a recommender system. We hope that our proposed design guidelines will serve as a helpful tool for researchers as well as practitioners in future.

It is important to know that in both user studies we view content and the visualization to be tightly interrelated in explanations as done in previous work. However, one can also plan experiments in which the effects of content and visualization are evaluated separately. Furthermore, the restriction to popular items in order to ensure high quality recommendations and/or tags is likely to have affected the rating distribution of the movies in both studies. This may have affected the positivity bias further, but was necessary to meet the minimum quality requirements for each interface. Another limitation results from the fact that the recommendation quality for each user could vary in both studies. By using popular items we tried to keep this limitation under control. Finally, note that our empirical studies only focused on the movie domain. According to [Tintarev and Masthoff, 2012], the movie domain suffers from being subjective in nature. However, their results show that their conclusions made on the movie domain could also be applied to the (more objective) domain of digital cameras.

7.2 Perspectives

One of the main problems of recommender systems is to develop and maintain a user model or user profile which is the system's representation of the user's preferences, interests, or characteristics. The user profile can either be acquired explicitly using, e.g., a 5-star rating scale, or derived automatically, e.g., by monitoring user behavior, in case no such explicit information is available. In the sign-up process of the movie recommender system MovieLens[1], for example, the system asks the user to rate at least 15 movies. These ratings help the system to build an initial version of the user profile which is necessary for computing personalized movie recommendations. The problem of missing data for computing recommendations is known as the *cold start and data sparsity* problem in recommender systems research. One way to solve this problem is to use additional user data which is already available on the Internet, e.g., on social networks such as Facebook or Twitter[2].

With the advent of the Social Web, user generated content has become the key to success for many of today's leading Web sites. User generated content is becoming more and more prevalent on the Internet, leading to the question of how to exploit data from the Social Web for recommender systems.

---

[1]http://www.movielens.org/join
[2]http://www.facebook.com, http://twitter.com

The main question addressed in this work was whether and how tagging data, which can be find on many Social Web sites, can be leveraged for the design of recommender systems. Note that the main focus of this thesis was not to alleviate the cold start problem by exploiting tagging data. However, we see this as one promising research direction.

As already stated in the introduction of this thesis, we made contributions on the fields of tag-based recommendations and tag-based explanations. We showed that leveraging tagging data is a valuable way to increase both the prediction accuracy of recommender algorithms as well as explanation quality. We believe that in future one of the main challenges will be to select the most relevant sources of knowledge in the Social Web to enhance existing recommender system techniques. We think that one day personalization techniques which filter the relevant from the irrelevant will be used to select the most appropriate input data from the overwhelming amount of Social Web data such as user location, friendship networks, or tagging data for recommender systems.

"We are leaving the age of information and entering the age of recommendation."
Chris Anderson in The Long Tail

Appendix A

Joint publications

Chapters 3, 4, 5, and 6 are partially based on publications, see [Kubatz et al., 2011], [Gedikli and Jannach, 2013], [Gedikli et al., 2011b], and [Gedikli et al., 2013][1] respectively. The main contributions of these research articles have been developed by the author of this thesis who is also the leading author of all papers except [Kubatz et al., 2011]. Details about the contributions of other researches and credits to students who supported this research work will be given in the following.

Chapter 3

The tag recommendation algorithm LocalRank is based on joint work with Marius Kubatz and Dietmar Jannach [Kubatz et al., 2011]. The main idea was contributed by the author of this thesis. In his diploma thesis, Marius Kubatz implemented the algorithm and conducted the experiments. The publication is based on many discussions between Marius Kubatz, Dietmar Jannach, and the author of this thesis.

Chapter 4

In [Gedikli and Jannach, 2013], Dietmar Jannach contributed with his discussions and support in implementing a baseline approach. The author of this thesis contributed the main idea and wrote the main part of the journal paper.

Chapter 5

The PHP-based Web application used in our study presented in [Gedikli et al., 2011b] was implemented by Martin Bring. Stefan Freudenreich conducted the experiment by recruiting participants and preparing the collected data. In their bachelor's theses, both Martin Bring and Stefan Freudenreich helped to realize a study which was designed and set-up by the author of this thesis, who also did the analysis and evaluation of the experiment results. The paper was jointly written by the authors Mouzhi Ge and Dietmar Jannach; the main technical parts were written by the author of the thesis.

Chapter 6

The Java-based Web application used in our second study [Gedikli et al., 2013] was implemented by Arash Baharloo in his bachelor thesis. Ayla Taşbaş, on the other hand, was responsible for recruiting participants from different user groups for the user study. In her diploma thesis, she also conducted the experiment and executed the SPSS tests. The study was designed and set-up by the author of this thesis, who also did the analysis and evaluation of the experiment results. The paper, which is currently being reviewed, was jointly written by the authors Dietmar Jannach and Mouzhi Ge; the main technical parts were written by the author of the thesis.

---

[1] This work is currently being reviewed.

Appendix B

Additional material for Chapter 6

B.1 Explanation interfaces used in the study

In this section, we give a short overview of the explanation interfaces evaluated in this study that were not described in the text so far. Besides a screenshot of each interface we provide a short description of the functionality. The user study was conducted in German language; for better readability, however, we include the English versions of the screenshots below.

B.1.1 Bar charts

Figure B.1 (a) shows a `barchart` explanation which basically contains a standard bar chart histogram showing the distribution of the target user's neighbors' ratings. There is one bar for each rating level (1-5).

According to the findings of [Herlocker et al., 2000], the `clusteredbarchart` explanation depicted in Figure B.1 (b), which clusters the "bad" ratings (1 and 2) and the "good" ratings (4 and 5) together, performs better because compared to the basic bar chart only a binary comparison between the good and bad ratings is required. This leads to a reduced decision effort for the user.

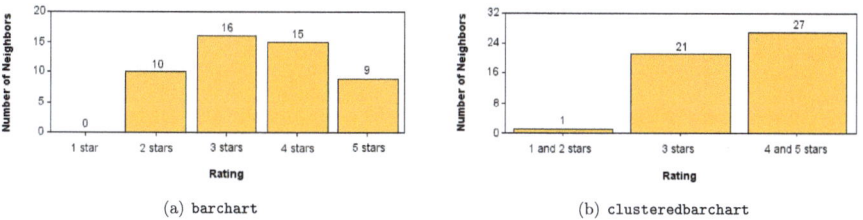

(a) barchart   (b) clusteredbarchart

Figure B.1: Bar chart explanations.

B.1.2 Rating- and prediction-based explanations

Figure B.2 (a) shows the explanation interface `average`, which presents the user with the overall average rating of the target item. The interface `confidence` shown in Figure B.2 (b) emulates Herlocker et al.'s *MovieLens percent confidence in prediction* interface. It shows a user's personalized prediction value together with the system's confidence in the prediction.

Figure B.3 (a) shows another interface which was evaluated in their user study. The `neighborsrating` explanation is a tabular view of the ratings within the user's neighborhood. The `piechart` explanation

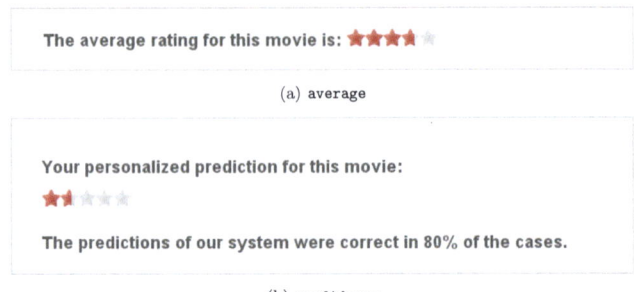

(a) average

(b) confidence

Figure B.2: Prediction-based explanations.

shown in Figure B.3 (b) represents the same data in a different way. It illustrates the distribution of the neighbors' ratings in a circle graph.

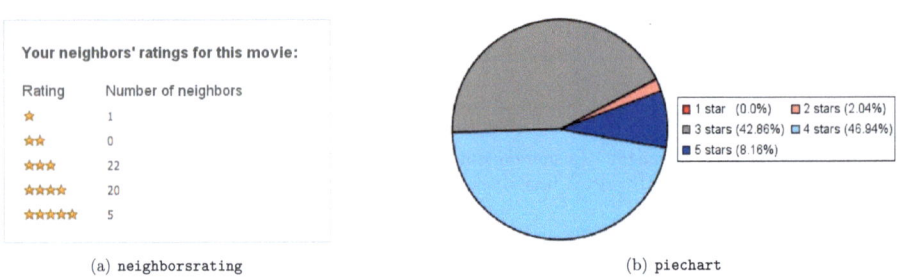

(a) neighborsrating

(b) piechart

Figure B.3: Rating-based explanations.

We also included two string-based explanation interfaces from [Herlocker et al., 2000]. The `neighborscount` interface shows the number of neighbors who provided a rating for the target item, whereas the `rated4+` explanation reveals the percentage of ratings for the target item which are equal or greater than 4.

In addition to the explanation interfaces presented so far, we included the newly proposed tag cloud interfaces in our study, as described in Section 6.2.2.

B.2 Statistics

B.2.1 Standard deviations

	efficiency	effectiveness	transparency	satisfaction
clusteredbarchart	10.02	2.07	1.25	1.60
barchart	10.37	1.90	1.26	1.40
neighborsrating	44.05	1.86	1.13	1.46
confidence	50.58	1.90	1.34	1.39
neighborscount	19.70	2.18	1.70	1.38
rated4+	10.65	1.96	1.28	1.50
average	10.13	1.77	1.46	1.39
tagcloud	77.75	1.57	1.60	1.91
perstagcloud	20.49	1.66	1.53	1.93
piechart	13.99	2.00	1.21	1.75

Table B.1: Standard deviation values.

B.2.2 Friedman test

	efficiency	effectiveness	transparency	satisfaction
clusteredbarchart	4.85	6.59	6.26	5.89
barchart	5.04	5.40	6.45	5.93
neighborsrating	5.36	5.38	5.45	4.81
confidence	4.89	4.84	4.62	5.70
neighborscount	5.66	3.23	2.20	2.21
rated4+	3.88	6.98	5.87	5.92
average	4.32	6.57	5.48	6.15
tagcloud	6.86	5.06	5.59	6.00
perstagcloud	7.37	4.83	6.96	6.87
piechart	6.76	6.11	6.13	5.52

Table B.2: Friedman test mean ranks.

	N	Chi-Square	df	Asymp. Sig.
efficiency	291	381.385	9	.000
effectiveness	291	415.036	9	.000
transparency	105	209.494	9	.000
satisfaction	105	180.849	9	.000

Table B.3: Friedman test statistics.

B.2.3 Wilcoxon Signed-Rank test

	efficiency		effectiveness		transparency		satisfaction	
	Z	Asymp. Sig.	Z	Asymp. Sig.	Z	Asymp. Sig.	Z	Asymp. Sig.
average-barchart	−2.956[a]	.003	−5.638[a]	.000	−2.600[a]	.009	−.644[a]	.520
average-clusteredbarchart	−2.056[a]	.040	−.059[b]	.953	−2.250[a]	.024	−.502[a]	.616
average-confidence	−2.430[a]	.015	−7.619[a]	.000	−2.442[b]	.015	−1.589[a]	.112
average-neighborscount	−5.577[a]	.000	−11.785[a]	.000	−8.104[b]	.000	−8.427[a]	.000
average-neighborsrating	−4.130[a]	.000	−5.959[a]	.000	−.264[a]	.792	−4.176[a]	.000
average-perstagcloud	−10.345[a]	.000	−7.788[a]	.000	−2.712[a]	.007	−1.290[b]	.197
average-piechart	−8.576[a]	.000	−2.605[a]	.009	−2.270[a]	.023	−1.853[a]	.064
average-rated4+	−1.810[b]	.070	−3.361[b]	.001	−1.809[a]	.070	−.339[a]	.735
average-tagcloud	−8.426[a]	.000	−6.298[a]	.000	−.015[a]	.988	−.275[a]	.784
barchart-clusteredbarchart	−.821[b]	.412	−5.872[b]	.000	−1.049[b]	.294	−.082[a]	.935
barchart-confidence	−.864[b]	.388	−3.310[a]	.001	−4.231[b]	.000	−.778[a]	.436
barchart-neighborscount	−2.850[a]	.004	−9.281[a]	.000	−8.050[a]	.000	−7.747[a]	.000
barchart-neighborsrating	−.911[a]	.362	−.834[a]	.404	−3.132[b]	.002	−3.485[a]	.000
barchart-perstagcloud	−8.253[a]	.000	−3.810[a]	.000	−.988[a]	.323	−2.346[b]	.019
barchart-piechart	−5.880[a]	.000	−3.185[b]	.001	−.728[b]	.466	−1.533[a]	.125
barchart-rated4+	−5.121[b]	.000	−7.123[b]	.000	−1.635[b]	.102	−.088[b]	.930
barchart-tagcloud	−6.666[a]	.000	−2.243[a]	.025	−2.753[b]	.006	−.322[b]	.747
clusteredbarchart-confidence	−.264[b]	.792	−7.417[a]	.000	−4.353[b]	.000	−.476[a]	.634
clusteredbarchart-neighborscount	−3.144[a]	.002	−10.370[a]	.000	−8.109[b]	.000	−7.679[a]	.000
clusteredbarchart-neighborsrating	−1.666[a]	.096	−6.408[a]	.000	−2.102[b]	.036	−3.021[a]	.003
clusteredbarchart-perstagcloud	−9.309[a]	.000	−6.876[a]	.000	−1.568[a]	.117	−1.994[b]	.046
clusteredbarchart-piechart	−6.753[a]	.000	−2.917[a]	.004	−.114[a]	.909	−1.408[a]	.159
clusteredbarchart-rated4+	−4.342[b]	.000	−2.172[b]	.030	−1.024[b]	.306	−.078[b]	.938
clusteredbarchart-tagcloud	−7.481[a]	.000	−5.336[a]	.000	−2.001[b]	.045	−.219[b]	.827
confidence-neighborscount	−3.612[a]	.000	−7.558[a]	.000	−6.746[b]	.000	−8.147[a]	.000
confidence-neighborsrating	−1.798[a]	.072	−2.988[b]	.003	−2.639[a]	.008	−3.067[a]	.002
confidence-perstagcloud	−9.907[a]	.000	−1.069[a]	.285	−4.763[a]	.000	−2.275[b]	.023
confidence-piechart	−6.683[a]	.000	−5.004[b]	.000	−4.113[a]	.000	−.546[a]	.585
confidence-rated4+	−4.765[b]	.000	−9.041[b]	.000	−3.498[a]	.000	−.794[b]	.427
confidence-tagcloud	−7.568[a]	.000	−.656[b]	.512	−1.982[a]	.048	−.435[b]	.663
neighborscount-neighborsrating	−2.445[b]	.014	−8.920[b]	.000	−7.980[a]	.000	−7.663[b]	.000
neighborscount-perstagcloud	−6.174[a]	.000	−4.922[b]	.000	−7.853[a]	.000	−7.666[b]	.000
neighborscount-piechart	−2.733[a]	.000	−9.896[b]	.000	−8.081[a]	.000	−7.558[b]	.000
neighborscount-rated4+	−6.741[b]	.000	−13.333[b]	.000	−8.294[a]	.000	−8.209[b]	.000
neighborscount-tagcloud	−3.515[a]	.000	−6.793[b]	.000	−7.183[a]	.000	−7.346[b]	.000
neighborsrating-perstagcloud	−8.295[a]	.000	−3.520[b]	.000	−2.712[a]	.007	−3.903[b]	.000
neighborsrating-piechart	−5.385[a]	.000	−3.308[b]	.001	−2.454[a]	.014	−2.059[b]	.040
neighborsrating-rated4+	−5.625[b]	.000	−7.487[b]	.000	−1.027[a]	.305	−3.107[b]	.002
neighborsrating-tagcloud	−5.904[a]	.000	−1.871[a]	.061	−.163[b]	.870	−2.485[b]	.013
perstagcloud-piechart	−4.870[b]	.000	−5.423[b]	.000	−1.301[b]	.193	−2.759[a]	.006
perstagcloud-rated4+	−11.895[b]	.000	−8.550[b]	.000	−2.000[b]	.045	−1.377[a]	.169
perstagcloud-tagcloud	−2.831[b]	.005	−2.333[b]	.020	−3.829[b]	.000	−2.323[a]	.020
piechart-rated4+	−10.006[b]	.000	−4.600[b]	.000	−1.028[b]	.304	−1.272[b]	.203
piechart-tagcloud	−2.216[a]	.027	−3.822[a]	.000	−1.968[b]	.049	−1.054[b]	.292
rated4+-tagcloud	−10.134[a]	.000	−8.146[a]	.000	−.981[b]	.327	−.097[b]	.923

Table B.4: Wilcoxon Signed-Rank test. a. Based on neg. ranks. b. Based on pos. ranks.

Bibliography

Adomavicius, G. and Kwon, Y. (2007). New recommendation techniques for multicriteria rating systems. *IEEE Intelligent Systems*, 22(3):48–55.

Adomavicius, G. and Tuzhilin, A. (2001). Extending recommender systems: A multidimensional approach. In *Proceedings of the Workshop on Intelligent Techniques for Web Personalization (ITWP'01)*, pages 4–6, Acapulco, Mexico.

Adomavicius, G. and Tuzhilin, A. (2005). Toward the next generation of recommender systems: A survey of the state-of-the-art and possible extensions. *IEEE Transactions on Knowledge and Data Engineering*, 17(6):734–749.

Agrawal, R. and Srikant, R. (1994). Fast algorithms for mining association rules in large databases. In *Proceedings of the 20th International Conference on Very Large Data Bases (VLDB'94)*, pages 487–499, Santiago de Chile, Chile.

Anand, S. S. and Mobasher, B. (2005). Intelligent Techniques for Web Personalization. In Mobasher, B. and Anand, S. S., editors, *Revised Selected Papers from the Workshop on Intelligent Techniques for Web Personalization (ITWP'03) at IJCAI'03*, volume 3169 of *LNAI*, pages 1–36. Springer.

Baeza-Yates, R. and Ribeiro-Neto, B. (1999). *Modern Information Retrieval.* Addison-Wesley.

Balabanovic, M. and Shoham, Y. (1997a). Combining content-based and collaborative recommendation. *Communications of the ACM*, 40(3):66–72.

Balabanovic, M. and Shoham, Y. (1997b). Fab: Content-based, collaborative recommendation. *Communications of the ACM*, 40(3):66–72.

Baur, D., Boring, S., and Butz, A. (2010). Rush: Repeated recommendations on mobile devices. In *Proceedings of the 15th International Conference on Intelligent User Interfaces (IUI'10)*, pages 91–100, Hong Kong, China.

Begelman, G., Keller, P., and Smadja, F. (2006). Automated tag clustering: Improving search and exploration in the tag space. In *Proceedings of the Collaborative Web Tagging Workshop at WWW'06*, Edinburgh, Scotland.

Bellogín, A., Cantador, I., and Castells, P. (2010). A study of heterogeneity in recommendations for a social music service. In *Proceedings of the 1st International Workshop on Information Heterogeneity and Fusion in Recommender Systems (HetRec'10)*, pages 1–8, Barcelona, Spain.

Berners-Lee, T. (2006). Linked data. http://www.w3.org/DesignIssues/LinkedData.html. Retrieved on July 11, 2010.

Berry, D. C. and Broadbent, D. E. (1987). Explanation and verbalization in a computer-assisted search task. *Quarterly Journal of Experimental Psychology*, 39(4):585–609.

Bilgic, M. and Mooney, R. J. (2005). Explaining recommendations: Satisfaction vs. promotion. In *Proceedings of the Workshop on the Next Stage of Recommender Systems Research (Beyond Personalization'05)*, pages 13–18, San Diego, CA, USA.

Billsus, D., Pazzani, M. J., and Chen, J. (2000). A learning agent for wireless news access. In *Proceedings of the 5th International Conference on Intelligent User Interfaces (IUI'00)*, pages 33–36, New Orleans, Louisiana, USA.

Bischoff, K., Firan, C. S., Nejdl, W., and Paiu, R. (2008). Can all tags be used for search? In *Proceedings of the 17th ACM Conference on Information and Knowledge Management (CIKM'08)*, pages 193–202, Napa Valley, CA, USA.

Bizer, C., Heath, T., Idehen, K., and Berners-Lee, T. (2008). Linked data on the Web. In *Proceedings of the 17th International Conference on World Wide Web (WWW'08)*, pages 1265–1266, Beijing, China.

Blei, D. M., Ng, A. Y., and Jordan, M. I. (2003). Latent dirichlet allocation. *Journal of Machine Learning Research*, 3:993–1022.

Bogers, T. and van den Bosch, A. (2009). Collaborative and content-based filtering for item recommendation on social bookmarking websites. In *Proceedings of the Workshop on Recommender Systems and the Social Web (RSWEB'09)*, pages 9–16, New York, NY, USA.

Breese, J. S., Heckerman, D., and Kadie, C. M. (1998). Empirical analysis of predictive algorithms for collaborative filtering. In *Proceedings of the 14th Conference on Uncertainty in Artificial Intelligence (UAI'98)*, pages 43–52, Madison, Wisconsin, USA.

Brin, S. and Page, L. (1998). The anatomy of a large-scale hypertextual Web search engine. *Computer Networks*, 30(1-7):107–117.

Brooks, C. H. and Montanez, N. (2006). Improved annotation of the blogosphere via autotagging and hierarchical clustering. In *Proceedings of the 15th International Conference on World Wide Web (WWW'06)*, pages 625–632, Edinburgh, Scotland.

Budanitsky, A. and Hirst, G. (2006). Evaluating WordNet-based measures of lexical semantic relatedness. *Computational Linguistics*, 32(1):13–47.

Bundschus, M., Yu, S., Tresp, V., Rettinger, A., Dejori, M., and Kriegel, H.-P. (2009). Hierarchical bayesian models for collaborative tagging systems. In *Proceedings of the IEEE International Conference on Data Mining (ICDM'09)*, pages 728–733, Miami, Florida, USA.

Cantador, I., Bellogín, A., and Vallet, D. (2010). Content-based recommendation in social tagging systems. In *Proceedings of the 2010 ACM Conference on Recommender Systems (RecSys'10)*, pages 237–240, Barcelona, Spain.

Cattuto, C., Benz, D., Hotho, A., and Stumme, G. (2008). Semantic grounding of tag relatedness in social bookmarking systems. In *Proceedings of the 7th International Conference on the Semantic Web (ISWC'08)*, pages 615–631, Karlsruhe, Germany.

Celma, O. and Herrera, P. (2008). A new approach to evaluating novel recommendations. In *Proceedings of the 2008 ACM Conference on Recommender Systems (RecSys'08)*, pages 179–186, Lausanne, Switzerland.

Chang, C.-C. and Lin, C.-J. (2011). Libsvm: A library for support vector machines. http://www.csie.ntu.edu.tw/~cjlin/libsvm. Retrieved on January 20, 2011.

Chen, L. and Pu, P. (2010). User evaluation framework of recommender systems. In *Proceedings of the Workshop on Social Recommender Systems (SRS'10)*, Hong Kong, China.

Chirita, P. A., Costache, S., Nejdl, W., and Handschuh, S. (2007). P-Tag: Large scale automatic generation of personalized annotation tags for the Web. In *Proceedings of the 16th International Conference on World Wide Web (WWW'07)*, pages 845–854, Banff, Alberta, Canada.

Cleverdon, C. and Kean, M. (1968). Factors determining the performance of indexing systems. In *Aslib Cranfield Research Project*, Cranfield, England.

Cosley, D., Lam, S. K., Albert, I., Konstan, J. A., and Riedl, J. T. (2003). Is seeing believing? How recommender system interfaces affect users' opinions. In *Proceedings of the SIGCHI Conference on Human Factors in Computing Systems (CHI'03)*, pages 585–592, Ft. Lauderdale, Florida, USA.

Cramer, H., Evers, V., Ramlal, S., Someren, M., Rutledge, L., Stash, N., Aroyo, L., and Wielinga, B. (2008). The effects of transparency on trust in and acceptance of a content-based art recommender. *User Modeling and User-Adapted Interaction*, 18(5):455–496.

Cremonesi, P., Garzotto, F., Negro, S., Papadopoulos, A., and Turrin, R. (2011). Comparative evaluation of recommender system quality. In *Proceedings of the 2011 Conference Extended Abstracts on Human Factors in Computing Systems (CHI EA'11)*, pages 1927–1932, Vancouver, BC, Canada.

de Gemmis, M., Lops, P., Semeraro, G., and Basile, P. (2008). Integrating tags in a semantic content-based recommender. In *Proceedings of the 2008 ACM Conference on Recommender Systems (RecSys'08)*, pages 163–170, Lausanne, Switzerland.

Demšar, J. (2006). Statistical comparisons of classifiers over multiple data sets. *Journal of Machine Learning Research*, 7:1–30.

Dias, M. B., Locher, D., Li, M., El-Deredy, W., and Lisboa, P. J. (2008). The value of personalised recommender systems to e-business: A case study. In *Proceedings of the 2008 ACM Conference on Recommender Systems (RecSys'08)*, pages 291–294, Lausanne, Switzerland.

Diederich, J. and Iofciu, T. (2006). Finding communities of practice from user profiles based on folksonomies. In *Proceedings of the 1st International Workshop on Building Technology Enhanced Learning Solutions for Communities of Practice (TEL-CoPs'06)*, pages 288–297, Crete, Greece.

Durao, F. and Dolog, P. (2009). Analysis of tag-based recommendation performance for a semantic wiki. In *Proceedings of the 4th Semantic Wiki Workshop (SemWiki'09)*, Hersonissos, Greece.

Durao, F. and Dolog, P. (2010). Extending a hybrid tag-based recommender system with personalization. In *Proceedings of the 2010 ACM Symposium on Applied Computing (SAC'10)*, pages 1723–1727, Sierre, Switzerland.

Ekstrand, M. D., Ludwig, M., Konstan, J. A., and Riedl, J. T. (2011). Rethinking the recommender research ecosystem: Reproducibility, openness, and lenskit. In *Proceedings of the 2011 ACM Conference on Recommender Systems (RecSys'11)*, pages 133–140, Chicago, Illinois, USA.

Felfernig, A. and Burke, R. (2008). Constraint-based recommender systems: Technologies and research issues. In *Proceedings of the ACM Conference on Electronic Commerce (EC'08)*, pages 1–10, Innsbruck, Austria.

Felfernig, A., Chen, L., and Mandl, M. (2011). Workshop on Human Decision Making in Recommender Systems (DECISION'11). In *Proceedings of the 2011 ACM Conference on Recommender Systems (RecSys'11)*, Chicago, Illinois, USA.

Felfernig, A., Friedrich, G., Jannach, D., and Zanker, M. (2007). An integrated environment for the development of knowledge-based recommender applications. *International Journal of Electronic Commerce*, 11(2):11–34.

Felfernig, A. and Gula, B. (2006). An empirical study on consumer behavior in the interaction with knowledge-based recommender applications. In *Proceedings of the IEEE Conference on E-Commerce Technology (CEC'06)*, pages 288–296, Washington, DC, USA.

Fellbaum, C., editor (1998). *WordNet: An Electronic Lexical Database (Language, Speech, and Communication)*. The MIT Press.

Firan, C. S., Nejdl, W., and Paiu, R. (2007). The benefit of using tag-based profiles. In *Proceedings of the 2007 Latin American Web Conference (LA-WEB'07)*, pages 32–41, Santiago de Chile, Chile.

Fleder, D. and Hosanagar, K. (2009). Blockbuster culture's next rise or fall: The impact of recommender systems on sales diversity. *Management Science*, 55(5):697–712.

Freyne, J., Anand, S. S., Guy, I., and Hotho, A. (2011). Workshop on Recommender Systems and the Social Web (RSWEB'11). In *Proceedings of the 2011 ACM Conference on Recommender Systems (RecSys'11)*, Chicago, Illinois, USA.

Friedrich, G. and Zanker, M. (2011). A taxonomy for generating explanations in recommender systems. *AI Magazine*, 32(3):90–98.

Funk, S. (2006). Netflix update: Try this at home. http://sifter.org/~simon/journal/20061211. html. Retrieved on January 20, 2011.

Ge, M., Jannach, D., Gedikli, F., and Hepp, M. (2012). Effects of the placement of diverse items in recommendation lists. In *Proceedings of the 14th International Conference on Enterprise Information Systems (ICEIS'12)*, pages 201–208, Wroclaw, Poland.

Gedikli, F., Bagdat, F., Ge, M., and Jannach, D. (2011a). RF-Rec: Fast and accurate computation of recommendations based on rating frequencies. In *Proceedings of the 13th IEEE Conference on Commerce and Enterprise Computing (CEC'11)*, pages 50–57, Luxembourg City, Luxembourg.

Gedikli, F., Ge, M., and Jannach, D. (2011b). Understanding recommendations by reading the clouds. In *Proceedings of the 12th International Conference on Electronic Commerce and Web Technologies (EC-Web'11)*, pages 196–208, Toulouse, France.

Gedikli, F. and Jannach, D. (2010a). Neighborhood-restricted mining and weighted application of association rules for recommenders. In *Proceedings of the 8th Workshop on Intelligent Techniques for Web Personalization & Recommender Systems (ITWP'10) at UMAP'10*, pages 8–19, Big Island, Hawaii.

Gedikli, F. and Jannach, D. (2010b). Neighborhood-restricted mining and weighted application of association rules for recommenders. In *Proceedings of the 11th International Conference on Web Information System Engineering (WISE'10)*, pages 157–165, Hong Kong, China.

Gedikli, F. and Jannach, D. (2010c). Rating items by rating tags. In *Proceedings of the 2nd Workshop on Recommender Systems and the Social Web (RSWEB'10)*, pages 25–32, Barcelona, Spain.

Gedikli, F. and Jannach, D. (2010d). Recommending based on rating frequencies. In *Proceedings of the 2010 ACM Conference on Recommender Systems (RecSys'10)*, pages 233–236, Barcelona, Spain.

Gedikli, F. and Jannach, D. (2010e). Recommending based on rating frequencies: Accurate enough? In *Proceedings of the 8th Workshop on Intelligent Techniques for Web Personalization & Recommender Systems (ITWP'10) at UMAP'10*, pages 65–70, Big Island, Hawaii.

Gedikli, F. and Jannach, D. (2013). Improving recommendation accuracy based on item-specific tag preferences. *ACM Transactions on Intelligent Systems and Technology*, 4(1).

Gedikli, F., Jannach, D., and Ge, M. (2013). An analysis of the effects of using different explanation styles in collaborative filtering recommender systems. Paper under review.

Gemmell, J., Ramezani, M., Schimoler, T., Christiansen, L., and Mobasher, B. (2009a). The impact of ambiguity and redundancy on tag recommendation in folksonomies. In *Proceedings of the 2009 ACM Conference on Recommender Systems (RecSys'09)*, pages 45–52, New York, NY, USA.

Gemmell, J., Schimoler, T., Mobasher, B., and Burke, R. (2010). Hybrid tag recommendation for social annotation systems. In *Proceedings of the 19th ACM International Conference on Information and Knowledge Management (CIKM'10)*, pages 829–838, Toronto, Canada.

Gemmell, J., Schimoler, T. R., Christiansen, L., and Mobasher, B. (2009b). Improving folkrank with item-based collaborative filtering. In *Proceedings of the Workshop on Recommender Systems and the Social Web (RSWEB'09)*, pages 17–24, New York, NY, USA.

Golbeck, J. (2009). Tutorial on using social trust for recommender systems. In *Proceedings of the 2009 ACM Conference on Recommender Systems (RecSys'09)*, pages 425–426, New York, NY, USA.

Goldberg, D., Nichols, D., Oki, B. M., and Terry, D. (1992). Using collaborative filtering to weave an information tapestry. *Communications of the ACM*, 35(12):61–70.

Golder, S. A. and Huberman, B. A. (2006). Usage patterns of collaborative tagging systems. *Journal of Information Science*, 32(2):198–208.

Good, N., Schafer, J. B., Konstan, J. A., Borchers, A., Sarwar, B., Herlocker, J. L., and Riedl, J. T. (1999). Combining collaborative filtering with personal agents for better recommendations. AAAI'99/IAAI'99, pages 439–446, Orlando, Florida, USA.

Greenwald, A. G. (1976). Within-subjects designs: To use or not to use. *Psychological Bulletin*, 83:216–229.

Guy, I., Chen, L., and Zhou, M. X. (2010a). Workshop on Social Recommender Systems (SRS'10). In *Proceedings of the 15th International Conference on Intelligent User Interfaces (IUI'10)*, Hong Kong, China.

Guy, I., Zwerdling, N., Ronen, I., Carmel, D., and Uziel, E. (2010b). Social media recommendation based on people and tags. In *Proceedings of the 33rd International ACM Conference on Research and Development in Information Retrieval (SIGIR'10)*, pages 194–201, Geneva, Switzerland.

Halvey, M. J. and Keane, M. T. (2007). An assessment of tag presentation techniques. In *Proceedings of the 16th International Conference on World Wide Web (WWW'07)*, pages 1313–1314, Banff, Alberta, Canada.

Hanani, U., Shapira, B., and Shoval, P. (2001). Information filtering: Overview of issues, research and systems. *User Modeling and User-Adapted Interaction*, 11(3):203–259.

Harter, S. P. (1996). Variations in relevance assessments and the measurement of retrieval effectiveness. *Journal of the American Society for Information Science*, 47(1):37–49.

Harvey, M., Baillie, M., Ruthven, I., and Carman, M. J. (2010). Tripartite hidden topic models for personalised tag suggestion. In *Proceedings of the 32nd European Conference on Advances in Information Retrieval (ECIR'10)*, pages 432–443, Milton Keynes, UK.

He, J. and Chu, W. W. (2010). A social network-based recommender system. *Data Mining for Social Network Data*, 12:47–74.

Heath, T. and Motta, E. (2007). Revyu.com: A reviewing and rating site for the Web of data. In *Proceedings of the 6th International and the 2nd Asian Semantic Web Conference (ISWC'07/ASWC'07)*, pages 895–902, Busan, Korea.

Hegelich, K. and Jannach, D. (2009). Effectiveness of different recommender algorithms in the mobile internet: A case study. In *Proceedings of the 7th Workshop on Intelligent Techniques for Web Personalization & Recommender Systems (ITWP'09) at IJCAI'09*, pages 41–50, Pasadena, CA, USA.

Heitmann, B. and Hayes, C. (2010). Using linked data to build open, collaborative recommender systems. In *AAAI Spring Symposium: Linked Data Meets Artificial Intelligence*, pages 76–81, Stanford University.

Herlocker, J. L., Konstan, J. A., Borchers, A., and Riedl, J. T. (1999). An algorithmic framework for performing collaborative filtering. In *Proceedings of the 22nd International ACM Conference on Research and Development in Information Retrieval (SIGIR'99)*, pages 230–237, Berkeley, CA, USA.

Herlocker, J. L., Konstan, J. A., and Riedl, J. T. (2000). Explaining collaborative filtering recommendations. In *Proceedings of the 2000 ACM Conference on Computer Supported Cooperative Work (CSCW'00)*, pages 241–250, Philadelphia, Pennsylvania, USA.

Herlocker, J. L., Konstan, J. A., and Riedl, J. T. (2002). An empirical analysis of design choices in neighborhood-based collaborative filtering algorithms. *Information Retrieval*, 5(4):287–310.

Herlocker, J. L., Konstan, J. A., Terveen, L. G., and Riedl, J. T. (2004). Evaluating collaborative filtering recommender systems. *ACM Transactions on Information Systems*, 22(1):5–53.

Hotho, A., Jäschke, R., Schmitz, C., and Stumme, G. (2006). Information retrieval in folksonomies: Search and ranking. In *Proceedings of the 3rd European Semantic Web Conference (ESWC'06)*, pages 411–426, Budva, Montenegro.

Hu, M., Lim, E.-P., and Jiang, J. (2010). A probabilistic approach to personalized tag recommendation. In *Proceedings of the 2nd IEEE International Conference on Social Computing (SocialCom'10)*, pages 33–40, Minneapolis, MN, USA.

Hu, R. (2012). *Design and user perception issues for personality-engaged recommender systems*. PhD thesis, Ecole Polytechnique Féedérale de Lausanne (EPFL).

Hu, R. and Pu, P. (2011). Enhancing collaborative filtering systems with personality information. In *Proceedings of the 2011 ACM Conference on Recommender Systems (RecSys'11)*, pages 197–204, Chicago, Illinois, USA.

Huang, J., Cheng, X.-Q., Guo, J., Shen, H.-W., and Yang, K. (2010). Social recommendation with interpersonal influence. In *Proceedings of the 19th European Conference on Artificial Intelligence (ECAI'10)*, pages 601–606, Amsterdam, The Netherlands.

Huang, Z., Chen, H., and Zeng, D. (2004). Applying associative retrieval techniques to alleviate the sparsity problem in collaborative filtering. *ACM Transactions on Information Systems*, 22(1):116–142.

Jameson, A. and Smyth, B. (2007). Recommendation to groups. In *The Adaptive Web: Methods and Strategies of Web Personalization*, pages 596–627. Springer, Berlin, Germany.

Jannach, D. and Hegelich, K. (2009). A case study on the effectiveness of recommendations in the mobile internet. In *Proceedings of the 2009 ACM Conference on Recommender Systems (RecSys'09)*, pages 41–50, New York, NY, USA.

Jannach, D., Karakaya, Z., and Gedikli, F. (2012). Accuracy improvements for multi-criteria recommender systems. In *Proceedings of the ACM Conference on Electronic Commerce (EC'12)*, pages 674–689, Valencia, Spain.

Jannach, D., Zanker, M., Felfernig, A., and Friedrich, G. (2010). *Recommender Systems - An Introduction*. Cambridge University Press.

Jäschke, R., Marinho, L. B., Hotho, A., Schmidt-Thieme, L., and Stumme, G. (2007). Tag recommendations in folksonomies. In *Proceedings of the 11th European Conference on Principles and Practice of Knowledge Discovery in Databases (PKDD'07)*, pages 506–514, Warsaw, Poland.

Jäschke, R., Marinho, L. B., Hotho, A., Schmidt-Thieme, L., and Stumme, G. (2008). Tag recommendations in social bookmarking systems. *AI Communications*, 21(4):231–247.

Ji, A.-T., Yeon, C., Kim, H.-N., and Jo, G.-S. (2007). Collaborative tagging in recommender systems. In *Proceedings of the 20th Australian Joint Conference on Artificial Intelligence (AUS-AI'07)*, pages 377–386, Gold Coast, Australia.

Jiang, J. J. and Conrath, D. W. (1997). Semantic similarity based on corpus statistics and lexical taxonomy. In *Proceedings of the International Conference on Research in Computational Linguistics*, pages 19–33, Taipei, Taiwan.

Kim, H.-N., Ji, A.-T., Ha, I., and Jo, G.-S. (2010a). Collaborative filtering based on collaborative tagging for enhancing the quality of recommendation. *Electronic Commerce Research and Applications*, 9(1):73–83.

Kim, W., Jeong, O.-R., and Lee, S.-W. (2010b). On social Web sites. *Information Systems*, 35(2):215–236.

Kiran, R. U. and Reddy, P. K. (2009). An improved multiple minimum support based approach to mine rare association rules. In *Proceedings of the IEEE Symposium on Computational Intelligence and Data Mining (CIDM'09)*, pages 340–347, Nashville, TN, USA.

Kohavi, R., Longbotham, R., Sommerfield, D., and Henne, R. M. (2009). Controlled experiments on the web: Survey and practical guide. *Data Mining and Knowledge Discovery*, 18(1):140–181.

Konstan, J. A., Miller, B. N., Maltz, D., Herlocker, J. L., Gordon, L. R., and Riedl, J. T. (1997). Grouplens: Applying collaborative filtering to usenet news. *Communications of the ACM*, 40(3):77–87.

Koren, Y. (2010). Factor in the neighbors: Scalable and accurate collaborative filtering. *ACM Transactions on Knowledge Discovery from Data*, 4(1):1:1–1:24.

Krestel, R., Fankhauser, P., and Nejdl, W. (2009). Latent dirichlet allocation for tag recommendation. In *Proceedings of the 2009 ACM Conference on Recommender Systems (RecSys'09)*, pages 61–68, New York, NY, USA.

Krishnan, V., Narayanashetty, P. K., Nathan, M., Davies, R. T., and Konstan, J. A. (2008). Who predicts better?: Results from an online study comparing humans and an online recommender system. In *Proceedings of the 2008 ACM Conference on Recommender Systems (RecSys'08)*, pages 211–218, Lausanne, Switzerland.

Kubatz, M., Gedikli, F., and Jannach, D. (2011). LocalRank - Neighborhood-based, fast computation of tag recommendations. In *Proceedings of the 12th International Conference on Electronic Commerce and Web Technologies (EC-Web'11)*, pages 258–269, Toulouse, France.

Lawrence, N. D. and Urtasun, R. (2009). Non-linear matrix factorization with gaussian processes. In *Proceedings of the 26th International Conference on Machine Learning (ICML'09)*, pages 601–608, Montréal, Québec, Canada.

Lemire, D. and Maclachlan, A. (2005). Slope one predictors for online rating-based collaborative filtering. In *Proceedings of the 5th SIAM International Conference on Data Mining (SDM'05)*, pages 471–480, Newport Beach, CA, USA.

Lewis, J. R. (1995). IBM computer usability satisfaction questionnaires: Psychometric evaluation and instructions for use. *International Journal of Human-Computer Interaction*, 7(1):57–78.

Li, Q. and Kim, B. M. (2003). An approach for combining content-based and collaborative filters. In *Proceedings of the 6th International Workshop on Information Retrieval with Asian Languages (AsianIR'03)*, pages 17–24, Sapporo, Japan.

Li, X., Guo, L., and Zhao, Y. E. (2008). Tag-based social interest discovery. In *Proceedings of the 17th International Conference on World Wide Web (WWW'08)*, pages 675–684, Beijing, China.

Liang, H., Xu, Y., Li, Y., and Nayak, R. (2008). Collaborative filtering recommender systems using tag information. In *Proceedings of the 2008 IEEE/WIC/ACM International Conference on Web Intelligence and Intelligent Agent Technology (WI-IAT'08)*, pages 59–62, Washington, DC, USA.

Liang, H., Xu, Y., Li, Y., and Nayak, R. (2009a). Collaborative filtering recommender systems based on popular tags. In *Proceedings of the 14th Australasian Document Computing Symposium (ADCS'09)*, University of New South Wales, Sydney, Australia.

Liang, H., Xu, Y., Li, Y., and Nayak, R. (2009b). Tag based collaborative filtering for recommender systems. In Wen, P., Li, Y., Polkowski, L., Yao, Y., Tsumoto, S., and Wang, G., editors, *Rough Sets and Knowledge Technology*, volume 5589 of *LNCS*, pages 666–673. Springer.

Linden, G. (2006a). Early Amazon: Shopping cart recommendations - Geeking with Greg. http://glinden.blogspot.com/2006/04/early-amazon-shopping-cart.html. Retrieved on April 13, 2012.

Linden, G. (2006b). Make data useful. http://www.powershow.com/view/97492-NTg2N/Greg_Linden_
flash_ppt_presentation. Retrieved on April 13, 2012.

Linden, G., Smith, B., and York, J. (2003). Amazon.com recommendations: Item-to-item collaborative
filtering. *IEEE Internet Computing*, 7(1):76–80.

Liu, N. N., Meng, X., Liu, C., and Yang, Q. (2011). Wisdom of the better few: Cold start recommendation via representative based rating elicitation. In *Proceedings of the 2011 ACM Conference on
Recommender Systems (RecSys'11)*, pages 37–44, Chicago, Illinois, USA.

Lohmann, S., Ziegler, J., and Tetzlaff, L. (2009). Comparison of tag cloud layouts: Task-related performance and visual exploration. In *Proceedings of the 12th IFIP TC 13 International Conference on
Human-Computer Interaction: Part I (INTERACT'09)*, pages 392–404, Uppsala, Sweden.

Luo, H., Fan, J., Keim, D. A., and Satoh, S. (2009). Personalized news video recommendation. In
Proceedings of the ACM International Conference on Multimedia (MM'09), pages 459–471, Beijing,
China.

Ma, H., Zhou, T. C., Lyu, M. R., and King, I. (2011). Improving recommender systems by incorporating
social contextual information. *ACM Transactions on Information Systems*, 29(2):9:1–9:23.

Marlin, B. M., Zemel, R. S., Roweis, S., and Slaney, M. (2007). Collaborative filtering and the missing
at random assumption. In *Proceedings of the 23rd Conference on Uncertainty in Artificial Intelligence
(UAI'07)*, pages 267–275, Vancouver, BC, Canada.

Massa, P. and Bhattacharjee, B. (2004). Using trust in recommender systems: An experimental analysis.
In *Proceedings of the 2nd International Conference on Trust Management (iTrust'04)*, pages 221–235,
Oxford, UK.

Mathes, A. (2004). Folksonomies - Cooperative classification and communication through shared meta-
data. Technical Report LIS590CMC (Doctoral Seminar), Graduate School of Library and Information
Science, University of Illinois Urbana-Champaign.

McCarthy, K., Reilly, J., McGinty, L., and Smyth, B. (2004). Thinking positively - Explanatory feedback
for conversational recommender systems. In *Proceedings of the European Conference on Case-Based
Reasoning (ECCBR'04)*, pages 115–124, Madrid, Spain.

McNee, S. M., Riedl, J. T., and Konstan, J. A. (2006). Being accurate is not enough: How accuracy
metrics have hurt recommender systems. In *Extended Abstracts on Human Factors in Computing
Systems (CHI'06)*, pages 1097–1101, Montréal, Québec, Canada.

McSherry, D. (2005). Explanation in recommender systems. *Artificial Intelligence Review*, 24(2):179–197.

Melville, P., Mooney, R. J., and Nagarajan, R. (2002). Content-boosted collaborative filtering for improved recommendations. In *Proceedings of the 18th National Conference on Artificial Intelligence
(AAAI'02)*, pages 187–192, Edmonton, Alberta, Canada.

Melville, P. and Sindhwani, V. (2010). Recommender systems. In *Encyclopedia of Machine Learning*,
pages 829–838.

Mika, P. (2007). Ontologies are us: A unified model of social networks and semantics. *Web Semantics:
Science, Services and Agents on the World Wide Web*, 5(1):5–15.

Miller, B. N., Konstan, J. A., and Riedl, J. T. (2004). Pocketlens: Toward a personal recommender
system. *ACM Transactions on Information Systems*, 22(3):437–476.

Mooney, R. J. and Roy, L. (2000). Content-based book recommending using learning for text categorization. In *Proceedings of the 5th ACM Conference on Digital Libraries (DL'00)*, pages 195–204, San
Antonio, Texas, USA.

Nakagawa, M. and Mobasher, B. (2003). A hybrid Web personalization model based on site connectivity. In *Proceedings of the Workshop on Web Mining and Web Usage Analysis (WebKDD'03)*, pages 59–70, Washington, DC, USA.

Nikolaeva, R. and Sriram, S. (2006). The moderating role of consumer and product characteristics on the value of customized on-line recommendations. *International Journal of Electronic Commerce*, 11(2):101–123.

Noll, M. G. and Meinel, C. (2007). Web search personalization via social bookmarking and tagging. In *Proceedings of the 6th International and the 2nd Asian Semantic Web Conference (ISWC'07/ASWC'07)*, pages 367–380, Busan, Korea.

O'Connor, M., Cosley, D., Konstan, J. A., and Riedl, J. T. (2001). Polylens: A recommender system for groups of users. In *Proceedings of the 7th Europeon Conference on Computer Supported Co-Operative Work (ECSCW'01)*, pages 199–218, Bonn, Germany.

O'Donovan, J. and Smyth, B. (2005a). Eliciting trust values from recommendation errors. In *Proceedings of the 18th International Florida Artificial Intelligence Research Society Conference (FLAIRS'05)*, pages 289–294, Clearwater Beach, Florida, USA.

O'Donovan, J. and Smyth, B. (2005b). Trust in recommender systems. In *Proceedings of the 10th International Conference on Intelligent User Interfaces (IUI'05)*, pages 167–174, San Diego, CA, USA.

Ong, L. S., Shepherd, B., Tong, L. C., Seow-Choen, F., Ho, Y. H., Tang, C. L., Ho, Y. S., and Tan, K. (1997). The colorectal cancer recurrence support (cares) system. *Artificial Intelligence in Medicine*, 11(3):175–188.

Passant, A. (2007). Using ontologies to strengthen folksonomies and enrich information retrieval in weblogs. In *Proceedings of the 1st International Conference on Weblogs and Social Media (ICWSM'07)*, Boulder, Colorado, USA.

Pazzani, M. J. and Billsus, D. (2007). Content-based recommendation systems. In Brusilovsky, P., Kobsa, A., and Nejdl, W., editors, *The Adaptive Web*, volume 4321 of *LNCS*, pages 325–341. Springer.

Peters, I. and Becker, P. (2009). *Folksonomies: Indexing and Retrieval in Web 2.0*. De Gruyter.

Pilá,szy, I. and Tikk, D. (2009). Recommending new movies: Even a few ratings are more valuable than metadata. In *Proceedings of the 2009 ACM Conference on Recommender Systems (RecSys'09)*, pages 93–100, New York, NY, USA.

Pitkow, J., Schütze, H., Cass, T., Cooley, R., Turnbull, D., Edmonds, A., Adar, E., and Breuel, T. (2002). Personalized search. *Communications of the ACM*, 45(9):50–55.

Popescul, A., Ungar, L. H., Pennock, D. M., and Lawrence, S. (2001). Probabilistic models for unified collaborative and content-based recommendation in sparse-data environments. In *Proceedings of the 17th Conference on Uncertainty in Artificial Intelligence (UAI'01)*, pages 437–444, San Francisco, CA, USA.

Porter, M. F. (1997). Readings in information retrieval. chapter An algorithm for suffix stripping, pages 313–316. Morgan Kaufmann Publishers Inc.

Pu, P. and Chen, L. (2006). Trust building with explanation interfaces. In *Proceedings of the 11th International Conference on Intelligent User Interfaces (IUI'06)*, pages 93–100, Sydney, Australia.

Pu, P. and Chen, L. (2007). Trust-inspiring explanation interfaces for recommender systems. *Knowledge-Based Systems*, 20(6):542–556.

Pu, P., Chen, L., and Hu, R. (2012). Evaluating recommender systems from the user's perspective: Survey of the state of the art. *User Modeling and User-Adapted Interaction*, 22(4–5):317–355.

Rendle, S., Balby Marinho, L., Nanopoulos, A., and Lars, S.-T. (2009). Learning optimal ranking with tensor factorization for tag recommendation. In *Proceedings of the 15th ACM International Conference on Knowledge Discovery and Data Mining (SIGKDD'09)*, pages 727–736, Paris, France.

Rendle, S. and Schmidt-Thie, L. (2010). Pairwise interaction tensor factorization for personalized tag recommendation. In *Proceedings of the 3rd ACM International Conference on Web Search and Data Mining (WSDM'10)*, pages 81–90, New York, NY, USA.

Resnick, P., Iacovou, N., Suchak, M., Bergstorm, P., and Riedl, J. T. (1994). Grouplens: An open architecture for collaborative filtering of netnews. In *Proceedings of the 1994 ACM Conference on Computer Supported Cooperative Work (CSCW'94)*, pages 175–186, Chapel Hill, North Carolina, USA.

Resnick, P. and Varian, H. R. (1997). Recommender systems. *Communications of the ACM*, 40(3):56–58.

Ricci, F. and Nguyen, Q. (2007). Acquiring and revising preferences in a critique-based mobile recommender system. *IEEE Intelligent Systems*, 22(3):22–29.

Ricci, F., Rokach, L., Shapira, B., and Kantor, P. B., editors (2011a). *Recommender Systems Handbook*. Springer.

Ricci, F., Semeraro, G., de Gemmis, M., and Lops, P. (2011b). Workshop on Decision Making and Recommendation Acceptance Issues in Recommender Systems (DEMRA'11). In *Proceedings of the 19th International Conference on Advances in User Modeling (UMAP'11)*, Girona, Spain.

Ronen, I., Shahar, E., Ur, S., Uziel, E., Yogev, S., Zwerdling, N., Carmel, D., Guy, I., Har'el, N., and Ofek-Koifman, S. (2009). Social networks and discovery in the enterprise. In *Proceedings of the 32nd International ACM Conference on Research and Development in Information Retrieval (SIGIR'09)*, pages 836–836, Boston, MA, USA.

Rowe, G. and Wright, G. (1993). Expert systems in insurance: A review and analysis. *International Journal of Intelligent Systems in Accounting, Finance & Management*, 2(2):129–145.

Salakhutdinov, R. and Mnih, A. (2008). *Advances in Neural Information Processing Systems*, chapter Probabilistic matrix factorization, pages 1257–1264. MIT Press.

Salton, G. (1989). *Automatic text processing: The transformation, analysis, and retrieval of information by computer.* Addison-Wesley Longman Publishing Co., Inc.

Salton, G., Wong, A., and Yang, C. (1975). A vector space model for information retrieval. *Journal of the American Society for Information Science*, 18(11):613–620.

Sandvig, J. J., Mobasher, B., and Burke, R. (2007). Robustness of collaborative recommendation based on association rule mining. In *Proceedings of the 2007 ACM Conference on Recommender Systems (RecSys'07)*, pages 105–112, Minneapolis, MN, USA.

Sarwar, B., Karypis, G., Konstan, J. A., and Riedl, J. T. (2000). Analysis of recommendation algorithms for e-commerce. In *Proceedings of the ACM Conference on Electronic Commerce (EC'00)*, pages 158–167, Minneapolis, MN, USA.

Sarwar, B., Karypis, G., Konstan, J. A., and Riedl, J. T. (2001). Item-based collaborative filtering recommendation algorithms. In *Proceedings of the 10th International Conference on World Wide Web (WWW'01)*, pages 285–295, Hong Kong, China.

Schein, A. I., Popescul, A., Ungar, L. H., and Pennock, D. M. (2002). Methods and metrics for cold-start recommendations. In *Proceedings of the 25th International ACM Conference on Research and Development in Information Retrieval (SIGIR'02)*, pages 253–260, Tampere, Finland.

Scott, J. P. (2000). *Social Network Analysis: A Handbook.* SAGE Publications.

Sen, S., Harper, F. M., LaPitz, A., and Riedl, J. T. (2007). The quest for quality tags. In *Proceedings of the 2007 International ACM Conference on Supporting Group Work (GROUP'07)*, pages 361–370, Sanibel Island, Florida, USA.

Sen, S., Vig, J., and Riedl, J. T. (2009a). Learning to recognize valuable tags. In *Proceedings of the 13th International Conference on Intelligent User Interfaces (IUI'09)*, pages 87–96, Sanibel Island, Florida, USA.

Sen, S., Vig, J., and Riedl, J. T. (2009b). Tagommenders: Connecting users to items through tags. In *Proceedings of the 18th International World Wide Web Conference (WWW'09)*, pages 671–680, Madrid, Spain.

Senecal, S. and Nantel, J. (2004). The influence of online product recommendations on consumers' online choices. *Journal of Retailing*, 80(2):159–169.

Seth, A. and Zhang, J. (2008). A social network based approach to personalized recommendation of participatory media content. In *Proceedings of the International Conferenece on Weblogs and Social Media (ICWSM'08)*, Seattle, USA.

Shang, S., Hui, P., Kulkarni, S. R., and Cuff, P. W. (2011). Wisdom of the crowd: Incorporating social influence in recommendation models. In *IEEE International Workshop on Hot Topics in Peer-to-Peer Computing and Online Social Networking (HotPost'11)*, Tainan, Taiwan.

Shani, G. and Gunawardana, A. (2011). Evaluating recommendation systems. In *Recommender Systems Handbook*, pages 257–297.

Shepitsen, A., Gemmell, J., Mobasher, B., and Burke, R. (2008). Personalized recommendation in social tagging systems using hierarchical clustering. In *Proceedings of the 2008 ACM Conference on Recommender Systems (RecSys'08)*, pages 259–266, Lausanne, Switzerland.

Shirky, C. (2005). Ontology is overrated. http://www.shirky.com/writings/ontology_overrated.html. Retrieved on July 11, 2010.

Sigurbjörnsson, B. and van Zwol, R. (2008). Flickr tag recommendation based on collective knowledge. In *Proceedings of the 17th International Conference on World Wide Web (WWW'08)*, pages 327–336, Beijing, China.

Smucker, M. D., Allan, J., and Carterette, B. (2007). A comparison of statistical significance tests for information retrieval evaluation. In *Proceedings of the 16th ACM Conference on Information and Knowledge Management (CIKM'07)*, pages 623–632, Lisbon, Portugal.

Soboroff, I. and Nicholas, C. (1999). Combining content and collaboration in text filtering. In *Proceedings of the Workshop on Machine Learning for Information Filtering at IJCAI'99*, pages 86–91, Stockholm, Sweden.

Song, Y., Zhuang, Z., Li, H., Zhao, Q., Li, J., Lee, W.-C., and Giles, C. L. (2008). Real-time automatic tag recommendation. In *Proceedings of the 31st International ACM Conference on Research and Development in Information Retrieval (SIGIR'08)*, pages 515–522, Singapore, Singapore.

Su, X. and Khoshgoftaar, T. M. (2009). A survey of collaborative filtering techniques. *Advances in Artificial Intelligence*, 2009:4:2–4:2.

Swearingen, K. and Sinha, R. (2002). Interaction design for recommender systems. In *Proceedings of the 4th Conference on Designing Interactive Systems (DIS'02)*, London, UK.

Symeonidis, P., Nanopoulos, A., and Manolopoulos, Y. (2008). Tag recommendations based on tensor dimensionality reduction. In *Proceedings of the 2008 ACM Conference on Recommender Systems (RecSys'08)*, pages 43–50, Lausanne, Switzerland.

Symeonidis, P., Nanopoulos, A., and Manolopoulos, Y. (2009). Moviexplain: A recommender system with explanations. In *Proceedings of the 2009 ACM Conference on Recommender Systems (RecSys'09)*, pages 317–320.

Symeonidis, P., Nanopoulos, A., and Manolopoulos, Y. (2010). A unified framework for providing recommendations in social tagging systems based on ternary semantic analysis. *IEEE Transactions on Knowledge and Data Engineering*, 22(2):179–192.

Thompson, C. A., Göker, M. H., and Langley, P. (2004). A personalized system for conversational recommendations. *Journal of Artificial Intelligence Research*, 21(1):393–428.

Ting, I.-H., Chang, P. S., and Wang, S.-L. (2012). Understanding microblog users for social recommendation based on social networks analysis. *Journal of Universal Computer Science*, 18(4):554–576.

Tintarev, N. and Masthoff, J. (2007a). Effective explanations of recommendations: User-centered design. In *Proceedings of the 2007 ACM Conference on Recommender Systems (RecSys'07)*, pages 153–156, Minneapolis, MN, USA.

Tintarev, N. and Masthoff, J. (2007b). A survey of explanations in recommender systems. In *Proceedings of the 2007 IEEE 23rd International Conference on Data Engineering Workshop (ICDEW'07)*, pages 801–810, Washington, DC, USA.

Tintarev, N. and Masthoff, J. (2008a). The effectiveness of personalized movie explanations: An experiment using commercial meta-data. In *Proceedings of the 5th International Conference on Adaptive Hypermedia and Adaptive Web-Based Systems (AH'08)*, pages 204–213, Hannover, Germany.

Tintarev, N. and Masthoff, J. (2008b). Over- and underestimation in different product domains. In *Workshop on Recommender Systems in Conjunction with the European Conference on Artificial Intelligence (ECAI'08)*, pages 14–19, Patras, Greece.

Tintarev, N. and Masthoff, J. (2011). Designing and evaluating explanations for recommender systems. In *Recommender Systems Handbook*, pages 479–510.

Tintarev, N. and Masthoff, J. (2012). Evaluating the effectiveness of explanations for recommender systems - Methodological issues and empirical studies on the impact of personalization. *User Modeling and User-Adapted Interaction*, 22(4–5):399–439.

Tso-Sutter, K. H. L., Marinho, L. B., and Schmidt-Thieme, L. (2008). Tag-aware recommender systems by fusion of collaborative filtering algorithms. In *Proceedings of the 2008 ACM Symposium on Applied Computing (SAC'08)*, pages 1995–1999, Fortaleza, Ceara, Brazil.

Tucker, L. (1966). Some mathematical notes on three-mode factor analysis. *Psychometrika*, 31(3):279–311.

Tuijnman, A. C. and Keeves, J. P. (1994). Path analysis and linear structural relations analysis. In *The International Encyclopaedia of Education*, pages 4229–4252. Oxford, Pergamon.

Turdakov, D. (2007). Recommender system based on user-generated content. In *Proceedings of the Colloquium on Databases and Information Systems (SYRCoDIS'07)*, Moscow, Russia.

Ulusoy, R. (2012). Utilisation of global tag information for localrank. Bachelor thesis, TU Dortmund.

Vatturi, P. K., Geyer, W., Dugan, C., Muller, M., and Brownholtz, B. (2008). Tag-based filtering for personalized bookmark recommendations. In *Proceedings of the 17th ACM Conference on Information and Knowledge Management (CIKM'08)*, pages 1395–1396, Napa Valley, CA, USA.

Vig, J., Sen, S., and Riedl, J. T. (2009). Tagsplanations: Explaining recommendations using tags. In *Proceedings of the 13th International Conference on Intelligent User Interfaces (IUI'09)*, pages 47–56, Sanibel Island, Florida, USA.

Vig, J., Soukup, M., Sen, S., and Riedl, J. T. (2010). Tag expression: Tagging with feeling. In *Proceedings of the 23nd ACM Symposium on User Interface Software and Technology (UIST'10)*, pages 323–332, New York, NY, USA.

Voorhees, E. M. (1993). Using WordNet to disambiguate word senses for text retrieval. In *Proceedings of the 16th International ACM Conference on Research and Development in Information Retrieval (SIGIR'93)*, pages 171–180, Pittsburgh, Pennsylvania, USA.

Wal, T. V. (2007). Folksonomy. http://vanderwal.net/folksonomy.html. Retrieved on April 27, 2012.

Wang, Z., Wang, Y., and Wu, H. (2010). Tags meet ratings: Improving collaborative filtering with tag-based neighborhood method. In *Proceedings of the Workshop on Social Recommender Systems (SRS'10)*, pages 15–23, Hong Kong, China.

Wildemuth, B. M. (2003). Why conduct user studies? The role of empirical evidence in improving the practice of librarianship. Presented at the 9th Conference on Professional Information Resources (INFORUM'03), Prague, Czech Republic.

Xu, G., Gu, Y., Dolog, P., Zhang, Y., and Kitsuregawa, M. (2011a). Semrec: A semantic enhancement framework for tag based recommendation. In *Proceedings of the 25th Conference on Artificial Intelligence (AAAI'11)*, pages 1267–1272, San Francisco, CA, USA.

Xu, G., Gu, Y., Zhang, Y., Yang, Z., and Kitsuregawa, M. (2011b). Toast: A topic-oriented tag-based recommender system. In *Proceedings of the 12th International Conference on Web Information System Engineering (WISE'11)*, pages 158–171, Sydney, Australia.

Yildirim, H. and Krishnamoorthy, M. S. (2008). A random walk method for alleviating the sparsity problem in collaborative filtering. In *Proceedings of the 2008 ACM Conference on Recommender Systems (RecSys'08)*, pages 131–138, Lausanne, Switzerland.

Yu, Z., Zhou, X., Hao, Y., and Gu, J. (2006). TV program recommendation for multiple viewers based on user profile merging. *User Modeling and User-Adapted Interaction*, 16(1):63–82.

Yuan, Q., Zhao, S., Chen, L., Liu, Y., Ding, S., Zhang, X., and Zheng, W. (2009). Augmenting collaborative recommender by fusing explicit social relationships. In *Proceedings of the Workshop on Recommender Systems and the Social Web (RSWEB'09)*, New York, NY, USA.

Zanardi, V. and Capra, L. (2008). Social ranking: Uncovering relevant content using tag-based recommender systems. In *Proceedings of the 2008 ACM Conference on Recommender Systems (RecSys'08)*, pages 51–58, Lausanne, Switzerland.

Zanardi, V. and Capra, L. (2011). A scalable tag-based recommender system for new users of the Social Web. In *Proceedings of the 22nd International Conference on Database and Expert Systems Applications - Volume Part I (DEXA'11)*, pages 542–557, Toulouse, France.

Zanker, M., Bricman, M., Gordea, S., Jannach, D., and Jessenitschnig, M. (2006). Persuasive online-selling in quality & taste domains. In *Proceedings of the 7th International Conference on Electronic Commerce and Web Technologies (EC-Web'06)*, pages 51–60, Krakow, Poland.

Zhang, J. and Pu, P. (2007). A recursive prediction algorithm for collaborative filtering recommender systems. In *Proceedings of the 2007 ACM Conference on Recommender Systems (RecSys'07)*, pages 57–64, Minneapolis, MN, USA.

Zhang, N., Zhang, Y., and Tang, J. (2009a). A tag recommendation system for folksonomy. In *Proceedings of the 2nd Workshop on Social Web Search and Mining (SWSM'09)*, pages 9–16, Hong Kong, China.

Zhang, Z.-K., Zhou, T., and Zhang, Y.-C. (2009b). Personalized recommendation via integrated diffusion on user-item-tag tripartite graphs. *The Computing Research Repository*, abs/0904.1989.

Zhao, S., Du, N., Nauerz, A., Zhang, X., Yuan, Q., and Fu, R. (2008). Improved recommendation based on collaborative tagging behaviors. In *Proceedings of the 2008 International Conference on Intelligent User Interfaces (IUI'08)*, pages 413–416, Gran Canaria, Canary Islands, Spain.

Zhen, Y., Li, W.-J., and Yeung, D.-Y. (2009). Tagicofi: Tag informed collaborative filtering. In *Proceedings of the 2009 ACM Conference on Recommender Systems (RecSys'09)*, pages 69–76, New York, NY, USA.

Zhou, T. C., Ma, H., King, I., and Lyu, M. R. (2009). Tagrec: Leveraging tagging wisdom for recommendation. In *Proceedings of the 12th IEEE International Conference on Computational Science and Engineering (CSE'09)*, pages 194–199, Vancouver, BC, Canada.

Ziegler, C.-N., Lausen, G., and Schmidt-Thieme, L. (2004). Taxonomy-driven computation of product recommendations. In *Proceedings of the 13th ACM International Conference on Information and Knowledge Management (CIKM'04)*, pages 406–415, Washington, DC, USA.

Ziegler, C.-N., McNee, S. M., Konstan, J. A., and Lausen, G. (2005). Improving recommendation lists through topic diversification. In *Proceedings of the 14th International Conference on World Wide Web (WWW'05)*, pages 22–32, Chiba, Japan.